MEDICAL PROCEDURES, TESTING AND TECHNOLOGY

GLUTATHIONE S-TRANSFERASES

STRUCTURE, FUNCTIONS AND CLINICAL ASPECTS

Medical Procedures, Testing and Technology

Additional books and e-books in this series can be found on Nova's website under the Series tab.

MEDICAL PROCEDURES, TESTING AND TECHNOLOGY

GLUTATHIONE S-TRANSFERASES

STRUCTURE, FUNCTIONS AND CLINICAL ASPECTS

IGOR AZEVEDO SILVA
EDITOR

Copyright © 2020 by Nova Science Publishers, Inc.

All rights reserved. No part of this book may be reproduced, stored in a retrieval system or transmitted in any form or by any means: electronic, electrostatic, magnetic, tape, mechanical photocopying, recording or otherwise without the written permission of the Publisher.

We have partnered with Copyright Clearance Center to make it easy for you to obtain permissions to reuse content from this publication. Simply navigate to this publication's page on Nova's website and locate the "Get Permission" button below the title description. This button is linked directly to the title's permission page on copyright.com. Alternatively, you can visit copyright.com and search by title, ISBN, or ISSN.

For further questions about using the service on copyright.com, please contact:
Copyright Clearance Center
Phone: +1-(978) 750-8400   Fax: +1-(978) 750-4470   E-mail: info@copyright.com.

NOTICE TO THE READER

The Publisher has taken reasonable care in the preparation of this book, but makes no expressed or implied warranty of any kind and assumes no responsibility for any errors or omissions. No liability is assumed for incidental or consequential damages in connection with or arising out of information contained in this book. The Publisher shall not be liable for any special, consequential, or exemplary damages resulting, in whole or in part, from the readers' use of, or reliance upon, this material. Any parts of this book based on government reports are so indicated and copyright is claimed for those parts to the extent applicable to compilations of such works.

Independent verification should be sought for any data, advice or recommendations contained in this book. In addition, no responsibility is assumed by the Publisher for any injury and/or damage to persons or property arising from any methods, products, instructions, ideas or otherwise contained in this publication.

This publication is designed to provide accurate and authoritative information with regard to the subject matter covered herein. It is sold with the clear understanding that the Publisher is not engaged in rendering legal or any other professional services. If legal or any other expert assistance is required, the services of a competent person should be sought. FROM A DECLARATION OF PARTICIPANTS JOINTLY ADOPTED BY A COMMITTEE OF THE AMERICAN BAR ASSOCIATION AND A COMMITTEE OF PUBLISHERS.

Additional color graphics may be available in the e-book version of this book.

Library of Congress Cataloging-in-Publication Data

Names: Silva, Igor Azevedo, editor. Title: Glutathione S-transferases : structure, functions and clinical aspects / Igor Azevedo Silva, editor.
Description: New York: Nova Science Publishers, [2020] | Series: Medical procedures, testing and technology | Includes bibliographical references and index. | Summary: "Glutathione S-transferase (GST) isoenzymes are an important superfamily of proteins that are present in homodimer or heterodimer structures. The isoenzymes included in this superfamily are involved in endobiotic and xenobiotic detoxification systems in animals, plants and microorganisms. This compilation focuses on the importance of functional aspects of polymorphic variants in different populations so the risk genotypes can be identified as indicators of disease susceptibility. Genetic studies will help to develop prognostic biomarkers for early prediction and risk assessment in patients, and enable clinicians to develop personalized treatment regimens. The authors examine the different kinds of reactive oxygen species and reactive species which act as carcinogens and cause damage to DNA, RNA, proteins and lipids, thereby resulting in human disease like bladder cancer. Also reviewed is the association of GST's common genetic variants with the development of infertility (failure to establish a clinical pregnancy after 12 months of regular, unprotected sexual intercourse) in women and in men"-- Provided by publisher. Identifiers: LCCN 2020028233 (print) | LCCN 2020028234 (ebook) | ISBN 9781536181883 (paperback) | ISBN 9781536182682 (adobe pdf)
Subjects: LCSH: Glutathione transferase. Classification: LCC QP606.G59 G57 2020 (print) | LCC QP606.G59 (ebook) | DDC 572/.792--dc23
LC record available at https://lccn.loc.gov/2020028233
LC ebook record available at https://lccn.loc.gov/2020028234

Published by Nova Science Publishers, Inc. † *New York*

Contents

Preface		**vii**
Chapter 1	An Overview of the Glutathione S-Transferase Family: Structure, Classifications and Clinical Implications *Sètondji Cocou Modeste Alexandre Yahouédéhou and Marilda Souza Goncalves*	1
Chapter 2	Genetics of Glutathione S-Transferases and Complex Diseases *Atar Singh Kushwah and Monisha Banerjee*	25
Chapter 3	Antioxidant Activity of Glutathione S-Transferases and Their Association with Bladder Cancer *Md. Bayejid Hosen and Yearul Kabir*	63
Chapter 4	Glutathione S-Transferases in Infertility *Maria Manuel Casteleiro Alves, António Hélio Oliani, Luiza Breitenfeld and Ana Cristina Ramalhinho*	95
Index		123

PREFACE

Glutathione S-transferase (GST) isoenzymes are an important superfamily of proteins that are present in homodimer or heterodimer structures. The isoenzymes included in this superfamily are involved in endobiotic and xenobiotic detoxification systems in animals, plants and microorganisms.

This compilation focuses on the importance of functional aspects of polymorphic variants in different populations so the risk genotypes can be identified as indicators of disease susceptibility. Genetic studies will help to develop prognostic biomarkers for early prediction and risk assessment in patients, and enable clinicians to develop personalized treatment regimens.

The authors examine the different kinds of reactive oxygen species and reactive species which act as carcinogens and cause damage to DNA, RNA, proteins and lipids, thereby resulting in human disease like bladder cancer.

Also reviewed is the association of GST's common genetic variants with the development of infertility (failure to establish a clinical pregnancy after 12 months of regular, unprotected sexual intercourse) in women and in men.

Chapter 1 - Glutathione S-transferase (GST) isoenzymes are an important superfamily of proteins that are present in homodimer or heterodimer structures. The isoenzymes included in this superfamily are involved in endobiotic and xenobiotic detoxification systems in animals, plants and microorganisms. These isoenzymes catalyze conjugation reactions between intermediate metabolites and glutathione (GSH), which protect cells from oxidative stress. Furthermore, some isoenzymes can act as chaperones or transporters, regulate nitric oxide and kinase signaling pathways, and participate in steroid synthesis, as well as isomerase and peroxidase activities. The importance of GSTs in clinical research has been demonstrated through various experimental models designed to target different classes of inhibitors, which can act as anthelmintic, antiparasitic, antimalarial or anticancer drugs. Some reports have also shown that cancer cells, as well as parasites, employ GST detoxification systems to escape the lethal effects of toxic metabolites derived from drugs. Finally, the absence of genes encoding GSTs, as well as polymorphisms linked to reductions in GST activity, have been associated with higher levels of carcinogens and an increased cancer risk, in addition to hepatotoxicity. Some *GST* genes have also been identified as biomarkers for susceptibility to allergens.

Chapter 2 - The glutathione-S-transferases (GSTs) are dimeric cytosolic xenobiotic-metabolizing enzymes that catalyze the conjugation of an active xenobiotic to GSH, an endogenous water-soluble substrate. The GST-mediated detoxification pathway ensures cellular protection from environmental insults and oxidative stress. Cellular ROS/RNS levels promote peroxidation of lipids and lipoproteins present in biomembranes, which leads to development of several pathological conditions. GSTs protect cells against deleterious actions of ROS/ RNS by promoting redox homeostasis through neutralization of excessive reactive metabolites, whose chemical actions elicit numerous signaling cascades associated with cell proliferation, inflammatory responses, apoptosis and senescence. Human GSTs are divided into three main families, namely, cytosolic (α-, β-, γ-, δ-, ξ-and ζ), mitochondrial (π) and membrane-bound (κ).These classes of GSTs originate from different chromosomes but share ~30% sequence identity and have cell specific distribution. Genetic variants in antioxidant enzyme encoding genes lead to decreased enzymatic activity leading to oxidative DNA damage and chronic inflammation. Moreover, genetic variants in *GSTM1, T1* and *P1* have been reported to be involved in development of type 2 diabetes mellitus (T2DM) and various cancers like cervical (CaCx), bladder, colon, gastrointestinal, lung, pulmonary diseases, neurological disorders and various pathological phenotypes. Interestingly, some of the common genetic variants like *GSTM1*-Del (16kb deletion), *GSTT1*-Del (54kbdeletion) polymorphisms (Null/Present) and *GSTP1* +313A/G (105I/V rs1695) are significantly associated with complex diseases *viz.* T2DM and CaCx in north Indian population as well as other ethnic groups. The *GST* genetic variants have also shown interesting association with treatment outcome in cervical cancer patients undergoing chemoradiotherapy *viz.* clinical response, vital status, overall survival and toxicity. In addition to the above, this chapter focuses on the importance of functional aspect of polymorphic variants in different populations so that the risk genotypes can be identified as indicators of disease susceptibility. Genetic studies will help to develop prognostic biomarkers for early prediction and risk assessment in patients and enable clinicians to develop personalized treatment regimes.

Chapter 3 - Glutathione S-transferases (GSTs) are phase II metabolic enzymes which participate in the cellular detoxification processes against xenobiotics and noxious compounds as well as against oxidative stress. According to cellular compartment, the GSTs are divided into three GSTs superfamilies. The cytoplasmic (cGSTs), mitochondrial (κGSTs), and microsomal (also known as MAPEG) GSTs. Among these families, cGSTs are the most complex and most linked to human diseases, and according to similarities in amino-acid sequences, different structure of genes, and immunological cross-reactivity, cGSTs are divided into seven subtypes (α, π, μ, θ, ω, σ, and δ). GSTs act as antioxidant by conjugating glutathione with toxic molecules and detoxify a wide range of endogenous and environmental reactive oxygen species (ROS) such as superoxide, nitric oxide, hydroxyl radicals, ethylene oxide, polycyclic aromatic hydrocarbon epoxide present in tobacco smoke and other reactive species (RS) such as hydrogen peroxide, peroxynitrite, hypochlorous acid. There are different kinds of ROS and RS which act as carcinogens and cause damage to DNA,

RNA, proteins and lipids and involved in the mutagenesis process, thereby resulting in human disease like bladder cancer. Bladder cancer is one of the most common cancers of the urinary tract which is said to be affected by occupational chemicals, for instance, tobacco, 2-naphthylamine, benzidine, and 4-aminobiphenyl etc. When cellular detoxifying enzymes are less active or malfunctioned, these chemicals escape the detoxifying reactions, and hence deposited in the bladder, thereby causing damage to bladder wall which initiate bladder tumorigenesis. It is evident from numerous studies that GSTs scavenge bladder tumor formation by detoxifying these chemicals and facilitating their excretion through urine without any damage to bladder epithelial cells. It is evidenced that single nucleotide polymorphisms of different GSTs superfamilies are associated with the occurrence of bladder cancer. Though the expression of GSTs is essential for eradication of toxic chemicals and tumor cells, overexpression of GSTs have also been associated with survival of tumor cells. Overexpression of GSTs cause neutralization of cellular ROS which are indispensable for apoptosis of tumor cells and detoxify anticancer drugs that are used for cancer treatment. It has also been shown that, some GSTs enzymes are exceptionally overexpressed in tumor cells and study of GSTs has introduced different anticancer drugs such as GST inhibitors, glutathione analogues, pro-drugs etc. and also new scaffolds and analogues are reported every year. This chapter will discuss all of these issues in depth focusing on the antioxidant activities of GSTs and their association with bladder cancer.

Chapter 4 - The Glutathione S-transferases (GSTs) family plays an important role in the detoxification of environmentally toxic compounds and products of oxidative stress, neutralizing ROS production. Oxidative stress is referred as the imbalance between oxidants and antioxidants and the generation of excessive amounts of reactive oxygen species (ROS). In a healthy body, ROS and antioxidants remain in balance. When the balance is disrupted towards an overabundance of ROS, oxidative stress occurs. The presence of unbalanced ROS can cause cellular damage and change cellular functions because they regulate protein activity and gene expression, which can lead to several effects. Oxidative stress is recognized to play a central role in the pathophysiology of many different disorders, including infertility. Infertility is a disease characterized by the failure to establish a clinical pregnancy after 12 months of regular, unprotected sexual intercourse or due to an impairment of a person's capacity to reproduce either as an individual or with his/her partner. The incidence of infertility differs between racial and ethnic groups because of the multifactorial nature of the disease. The etiology and pathogenesis of infertility are still unclear. However, there is increasing evidence that infertility depends on complex interactions between genetic factors and environmental toxins which can be implicated in its pathogenesis. Because of GSTs detoxification properties, it is logical to suspect that dysfunction of detoxification-related enzymes might be a contributor to the development of infertility. In this chapter, the authors will review the association of GSTs common genetic variants with the development of infertility in women and in men.

In: Glutathione S-Transferases
Editor: Igor Azevedo Silva

ISBN: 978-1-53618-188-3
© 2020 Nova Science Publishers, Inc.

Chapter 1

AN OVERVIEW OF THE GLUTATHIONE S-TRANSFERASE FAMILY: STRUCTURE, CLASSIFICATIONS AND CLINICAL IMPLICATIONS

Sètondji Cocou Modeste Alexandre Yahouédéhou and Marilda Souza Goncalves, PhD*

Instituto Gonçalo Moniz, Salvador, Bahia, Brasil

ABSTRACT

Glutathione S-transferase (GST) isoenzymes are an important superfamily of proteins that are present in homodimer or heterodimer structures. The isoenzymes included in this superfamily are involved in endobiotic and xenobiotic detoxification systems in animals, plants and microorganisms. These isoenzymes catalyze conjugation reactions between intermediate metabolites and glutathione (GSH), which protect cells from oxidative stress. Furthermore, some isoenzymes can act as

---

* Corresponding Author's E-mail: mari@bahia.fiocruz.br.

chaperones or transporters, regulate nitric oxide and kinase signaling pathways, and participate in steroid synthesis, as well as isomerase and peroxidase activities. The importance of GSTs in clinical research has been demonstrated through various experimental models designed to target different classes of inhibitors, which can act as anthelmintic, antiparasitic, antimalarial or anticancer drugs. Some reports have also shown that cancer cells, as well as parasites, employ GST detoxification systems to escape the lethal effects of toxic metabolites derived from drugs. Finally, the absence of genes encoding GSTs, as well as polymorphisms linked to reductions in GST activity, have been associated with higher levels of carcinogens and an increased cancer risk, in addition to hepatotoxicity. Some *GST* genes have also been identified as biomarkers for susceptibility to allergens.

Keywords: *GST*, polymorphism, cancer, parasite, allergy, chemotherapy, resistance

INTRODUCTION

In general, the metabolism of xenobiotics and endobiotics occurs in two distinct phases (phases I and II), involving the participation of both import and export transporters. Phase I metabolism involves nonsynthetic reactions, such as oxidation, reduction, and hydrolysis, while the synthetic reactions that occur in phase II metabolism involve the conjugation of substrates with endogenous molecules, such as glucuronic acid, sulfate, acetyls, methyls and glutathione (GSH) (Božina, Bradamante, and Lovrić 2009; Bock 2014). Both synthetic and nonsynthetic reactions are catalyzed by enzymes, including cytochrome P450 (CYP450), myeloperoxidases (MPO), hydrolases, oxidases, epoxygenases reductases, hydroxylases, dehydrogenases, UDP-glucuronosyltransferases, sulfotransferases, N-acetyltransferases, methyltransferases and glutathione S-transferases (GST) (Božina, Bradamante, and Lovrić 2009).

The GSTs are a superfamily of isoenzymes crucial to xenobiotic and endobiotic detoxification in animals, plants and microorganisms (Martínez-Guzmán et al. 2017; Lawless et al. 2018). These ubiquitous isoenzymes are involved in phase II metabolism (Zhou et al. 2014), and are the principal catalyst in the conjugation of electrophilic compounds with nucleophilic GSH, a tripeptide consisting of cysteine, glycine and glutamic acid (Al-Qattan, Mordi, and Mansor 2016; Gulçin et al. 2018). GSTs are mainly expressed in the liver, but also in other tissues, e.g., the kidney, retina, lung, eye lens, cornea and brain, as well as erythrocytes and the placenta (Gulçin et al. 2018).

Due to their importance in the biological processes of normal and malign cells, as well as microorganisms, various studies have aimed to identify biomarkers in the genes encoding GST isoenzymes. These biomarkers hold promise for use in diagnostics and/or prognosis in some diseases, and may be useful for drug discovery and design. This chapter will present a brief overview of the structure and classifications of GSTs, with a primary focus on the main isoenzymes associated with clinical implications.

STRUCTURE AND CLASSIFICATION

Human GSTs are divided into three main families: cytosolic, mitochondrial and microsomal. The former two are soluble enzymes, while the latter are membrane-associated proteins in eicosanoid and glutathione metabolism (MAPEG) (Martínez-Guzmán et al. 2017). Protein structure analysis has shown that microsomal and cytosolic GSTs are distinct; however, both present functionally similar activity (McIlwain, Townsend, and Tew 2006).

Human erythrocyte GST consists of two subunits with a molecular weight of 47.5kDa (Marcus, Habig, and Jakoby 1978; Ricci et al. 1989). Cytosolic GSTs present as homodimers or heterodimers, with the molecular weight of subunits ranging from 24.5 to 28.5kDa (Martínez-Guzmán et al. 2017). The cytosolic GSTs are divided into eight classes: alpha, kappa, mu, omega, pi, sigma, theta and zeta (Lawless et al. 2018), which are encoded by *GSTA*, *GSTK*, *GSTM*, *GSTO*, *GSTP*, *GSTS*, *GSTT* and *GSTZ* genes, respectively (Safarinejad, Shafiei, and Safarinejad 2011).

These classifications take into account similarities in the primary structures of each isoform, as well as substrate specificities (Chikezie 2011).

The isoenzymes included within each class share more than 60% similarity, principally in the N-terminal domain. Studies have identified two important sites in the structure of GST proteins: the glutathione-site (G-site) and hydrophobic-site (H-site) (Al-Qattan, Mordi, and Mansor 2016). The substrate-binding H-site is proximal to the G-site, which is the catalytic site of the enzyme. The G-site interacts with the thiol group of GSH, catalyzing the conjugation of the substrate at the H-site (McIlwain, Townsend, and Tew 2006). It has been demonstrated that differences in H-site structure are related to isoenzymes, which define substrate selectivity (Martos-Maldonado et al. 2012). As the purification and characterization of GST from *Wuchereria bancrofti* revealed a putative non-catalytic binding site in its protein structure (Rajaiah Prabhu et al. 2018), it was suggested that this site could facilitate the fixation of endogenous and exogenous substrates.

Structural analysis of the human GSTA1-1 isoform indicates that it consists of 222 amino acid residues, with two domains: the N-terminal and C-terminal domains. The former (residues 3-83) is a thioredoxin site, while the latter (residues 85-207) contains five α-helices. This is followed by residues 210-220, named helix α9, whose conformation depends on ligand binding (Lawless et al. 2018).

FUNCTIONS

As previously mentioned, GSTs mainly catalyze the formation of a thioester bond between the intermediate metabolite and the GSH sulfur atom (Gulçin et al. 2018), which facilitates the elimination of toxic metabolites. Indeed, the formation of the GS-substrate complex increases the solubility of the intermediate metabolite, which is generally hydrophobic (Martínez-Guzmán et al. 2017).

GSTs detoxify a wide variety of substrates, including chemical carcinogens, antineoplastic agents and environmental toxins (Zhou et al. 2014; Martínez-Guzmán et al. 2017). These enzymes are also involved in the inactivation of oxidative stress products, such as α and β-unsaturated aldehydes, quinones, epoxides and hydroperoxides (Zhou et al. 2014). Studies have shown that GSTs catalyze the detoxification of both reactive metabolites derived from drugs, e.g., clozapine, diclofenac, mefenamic acid and nevirapine (Zhang et al. 2017), as well as oxidative metabolites derived from industrial chemicals, such as ethylene epoxide, methyl chloride and methyl bromide, which are potential inducers of prostate cancer (Gsur et al. 2001). Thus, GSTs can prevent cellular cytotoxicity and mutagenicity (Zhang et al. 2017), and may be involved in prodrug activation (Lawless et al. 2018).

In addition to catalytic activity, GSTs also exert non-catalytic functions (Al-Qattan, Mordi, and Mansor 2016), including the modulation of signaling pathways of cell proliferation and differentiation, apoptosis, intracellular transport and the biosynthesis of leukotrienes, prostaglandins, testosterone and progesterone (Zhou et al. 2014). Moreover, GSTs, for example hGSTA1, also catalyze GSH-dependent steroid isomerase and selenium-independent peroxidase reactions (Coles and Kadlubar 2005).

CLINICAL ASPECTS

GST genes are polymorphic, and carriers of polymorphisms associated with reduced or absent catalytic activity are susceptible to increased sensitivity to toxic compounds, and therefore enhanced disease susceptibility (Zhou et al. 2014). Some isoenzymes among the GST classes have been extensively investigated in several populations, revealing implications in the physiopathology of some diseases, as well as drug resistance in human cells and microorganisms.

Associations between GSTs and Disease Pathogenesis

GSTA

In humans, GST alpha class (hGSTA), also known as ligandin, consists of four isoforms: hGSTA1-hGSTA4 (Koumaravelou et al. 2011). hGSTA1 is the most abundant GST encountered in the liver, catalyzing the metabolism of a wide range of substrates, including carcinogenic compounds, chemotherapeutic agents and lipid peroxidation products (Martínez-Guzmán et al. 2017). The participation of hGSTA1 in the metabolism and intracellular regulation of 4-hydroxy-2-trans-nonenal, a lipid peroxidation product, determines whether cells will proliferate or initiate apoptosis. Martinez-Guzmán and colleagues demonstrated that the activation of glucocorticoid receptors induces an increase in hGSTA1 gene expression and activity, leading to the metabolism of its substrate, as well as the inhibition of cisplatin-induced apoptosis (Martínez-Guzmán et al. 2017). Moreover, hGSTA1 catalyzes the conjugation reaction of GSH with diol epoxides, the products of phase I metabolism of benzo [a]pyrene (BaP), a carcinogenic agent. Thus, glucocorticoids may inactivate the intermediate metabolites derived from BaP, which is associated with a reduction in DNA-adduct formation (Sundberg et al. 2002; Martínez-Guzmán et al. 2017).

The h*GSTA1* gene, located on chromosome 6 (6p12.1), has two alleles, *GSTA1**A and *GSTA1**B. The *GSTA1**A allele is defined by polymorphisms -567T, -69C and -52G, while *GSTA1**B is defined by -567G, -69T and -52A (Koumaravelou et al. 2011). Polymorphism -G52A has been linked to lower GSTA1 expression by preventing transcription factor Sp1 from binding to the promoter region (Koumaravelou et al. 2011). The *GSTA1* A/B and B/B genotypes are associated with reduced GSTA1 expression (Sá et al. 2014). While polymorphisms linked to low hGSTA1 expression have been associated with increased cancer risk (Sergentanis and Economopoulos 2010), a study conducted in a Brazilian population showed that individuals with *GSTA1* A/B and B/B genotypes presented a lower risk of prostate cancer compared to carriers of *GSTA1* A/A (Sá et al. 2014). Furthermore, an investigation conducted in the

Vietnamese population demonstrated the association of the *GSTA1* A/A genotype with gastric cancer risk (Nguyen et al. 2009). Ahn and colleagues found that female carriers of *GSTA1* A/B and B/B genotypes who smoked or consumed less cruciferous vegetables were at increased risk of developing breast cancer (Ahn et al. 2006). Chronic inflammatory conditions, such as oxidative stress and proliferative hyperplasia, were linked to increased levels of hGSTA1, while decreased levels were associated with prostate cancer (Sá et al. 2014).

Den Braver and colleagues demonstrated that susceptibility to the diclofenac-induced liver injury observed in some individuals may vary due to differences in GST expression. Furthermore, individuals with low hepatic GSTA1 and GSTM1 activity and elevated CYP activity may present an increased risk of diclofenac-induced liver injury (den Braver et al. 2016). In a recent study, Zhang and colleagues suggested that reduced GSTA1-1 and GSTA2-2 expression, in combination with increased CYP or MPO expression, may lead to increased susceptibility to amodiaquine, resulting in idiosyncratic hepatotoxicity and agranulocytosis in malaria patients (Zhang et al. 2017). Indeed, as CYP and MPO are involved in phase I metabolism, the increased expression of the genes encoding these enzymes can lead to higher levels of toxic metabolites in individuals presenting low GST expression.

GSTM and GSTT

The GSTM class consists of five isoenzymes, GSTM1-M5, encoded by *GSTM1-M5* genes located on chromosome 1 (1p13.3) (Jansson et al. 2003; Kabesch et al. 2004; Barjui, Reiisi, and Bayati 2017). The *GSTM1* gene has three alleles, *GSTM1* null (*GSTM1*^{0}) characterized by a complete deletion of the gene, as well as *GSTM1*^{A} and *^{B}, which are characterized by amino acid substitutions (Jansson et al. 2003). *GSTM1*^{A} differs from *GSTM1*^{B} in the G534C mutation, characterized by the substitution of asparagine (Asn) with lysine (Lys) (Sá et al. 2014). GSTM1 is expressed in the liver, stomach, brain and breast, and is involved in the detoxification of polycyclic aromatic hydrocarbons and oxidative stress products (Barjui, Reiisi, and Bayati 2017). The GSTT class consists of two isoenzymes,

GSTT1 and T2, encoded by *GSTT1* and *T2* genes, which are located on chromosome 22 (22q11.2) (Allan et al. 2001; Barjui, Reiisi, and Bayati 2017). GSTT1, which is expressed in the liver and erythrocytes, participates in the detoxification of several compounds, including ethylene oxide, methyl bromide, alkyl halides, benzo, acrolein, small hydrocarbons and halogenated metabolites (Zhou et al. 2014; Barjui, Reiisi, and Bayati 2017).

The complete deletion of both *GSTM1* and *GSTT1* genes results in the lack of enzyme activity (Safarinejad, Shafiei, and Safarinejad 2011). In combination, *GSTM1* and *T1* null genotypes have been associated with a worse prognosis in ovarian cancer (Abbas et al. 2015). In addition, the *GSTM1* null genotype was linked to a significantly reduced recurrence of cervical cancer. Stanulla and colleagues also demonstrated that children with acute lymphoblastic leukemia who carried both *GSTM1* and *T1* null genotypes, as well as the *GSTP1* -313GG variant genotype, had a lower risk of recurrence (Stanulla et al. 2000). The *GSTM1* and *T1* null genotypes have been associated with prostate cancer risk in Caucasian and Japanese populations, respectively (Safarinejad, Shafiei, and Safarinejad 2011). Moreover, individuals carriers of both *GSTM1* and *T1* null genotypes and the *GSTP1* Val allele were found to be at increased risk of prostate cancer, corroborating results from a study performed in the Indian population (Srivastava et al. 2005; Safarinejad, Shafiei, and Safarinejad 2011).

In some cases, patients with prostate cancer submitted to surgery or radiotherapy may present increases in serum levels of prostate-specific antigen, defined as biochemical recurrence, indicating the reappearance of cancer (Chen et al. 2013; Tourinho-Barbosa et al. 2018). The *GSTM1* null genotype and *GSTP1* -A313G polymorphism, in addition to *GSTP1* CpG hypermethylation, may be used as biomarkers to detect the biochemical recurrence of prostate cancer following radical prostatectomy (Chen et al. 2013). Nock and colleagues observed, in African-American men with high-grade and high-stage prostate tumors, a strong association between the *GSTT1* null genotype and an elevated risk of biochemical recurrence. Furthermore, Caucasians with high-grade and high-stage tumors who were carriers of the *GSTM1* null genotype presented an increase in biochemical recurrence (Nock et al. 2009). These authors suggested that GST-targeting agents might hold promise with regard to treatment outcome in African-Americans with advanced prostate cancers. Other authors have demonstrated an association between higher levels of anti-GSTM1 antibody and an increased risk of glaucoma, which may be due to higher GSTM1 levels present in these individuals (Yang et al. 2001). However, a study performed in the Swedish population found no associations between the functional *GSTM1* gene and glaucoma risk (Jansson et al. 2003).

Studies performed in different populations identified a higher risk of chronic renal failure in association with the *GSTM1* and *T1* null genotypes and *GSTP1* 105Val allele variant (Akgul et al. 2012). Singh and colleagues discovered that the *GSTM1* null genotype was linked to an increased risk of renal transplant rejection, while the *GSTP1* 105Val allele variant was associated with delayed graft function (Singh et al. 2009). Contrarily, Pagliuso and colleagues found no associations between *GSTM1* and *T1* polymorphisms and chronic allograft nephropathy (Pagliuso et al. 2008). It was suggested that renal allograft dysfunction may occur in recipient carriers of the *GSTT1* null genotype due to anti-GSTT1 antibody production against GSTT1 expressed on the donor graft (Akgul et al. 2012). However, it was also shown that both recipient and donor carriers of the *GSTT1* null genotype produce anti-GSTT1 antibody against GSTT1 expressed on liver grafts (Wichmann et al. 2006), perhaps as a result of previous blood transfusion, pregnancy or transplantation. Anti-GSTT1 production was estimated to occur at approximately 43.6 months after transplantation, followed by the development of chronic antibody-mediated rejection, which occurs at 65.4 months after transplantation, i.e., 21.8 months after the production of antibodies (Akgul et al. 2012).

Compared to anti-human leukocyte antigen antibodies, the anti-GSTT1 antibody induces tissue damage more slowly (Akgul et al. 2012). Performing anti-GSTT1 antibody screening in patients submitted to liver transplantation, Salcedo and colleagues identified, at early stages, those with higher risk of *de novo* autoimmune hepatitis (Salcedo et al. 2009). They suggested that the anti-GSTT1 antibody could be used as a biomarker of rejection risk in post-renal and liver transplantation, and demonstrated the importance of including the *GSTT1* null genotype when investigating immunological parameters in transplant candidates (Akgul et al. 2012).

It was suggested that reduced or lacking GSTT1 and GSTM1 activity could be a risk factor for developing neuropathic and neurodegenerative diseases (Barjui, Reiisi, and Bayati 2017). Indeed, the reduction or absence of GST activity leads to decreases in GSH metabolism and, consequently, an increase in reactive oxygen species-induced injury in nerve cells. A study performed in the Iranian population demonstrated the association of *GSTM1* and *T1* null genotypes with increased risk of multiple sclerosis, a chronic inflammatory disease (Barjui, Reiisi, and Bayati 2017). Kabesh and colleagues demonstrated that individuals with *GSTM1* and *GSTT1* null genotypes are at higher risk of developing asthma and related symptoms (Kabesch et al. 2004). These authors suggested that this may result from in utero or current exposure to environmental tobacco smoke in combination with the reduction or lack of GSTs or other enzymes involved in toxic metabolite detoxification, thereby contributing to side effects associated with passive and active tobacco smoking. Corroborating their hypothesis, another study demonstrated the effects of smoking during pregnancy on fetus growth and fetal lung maturation (Cook and Strachan 1999). Moreover, reports have shown that individual carriers of the *GSTM1* null genotype present increased levels of aromatic DNA adducts in lung tissue and cytogenetic alterations in lung cells exposed to smoke (van Poppel et al. 1992; Ryberg et al. 1997). The *GSTT1* null genotype has also been associated with exacerbated DNA damage (Kabesch et al. 2004). The *GSTM1* and *T1* null genotypes were also found to be associated with higher idiosyncratic drug-induced liver injury risk related to nonsteroidal anti-inflammatory drugs (den Braver et al. 2016).

GSTP

Human GSTP is encoded by the *GSTP1* gene located on chromosome 11 (11q13) (Oh et al. 2005). This isoenzyme is most abundant in the lung epithelium and is also found in the placenta, breast and prostate (Oh et al. 2005; López-Rodríguez et al. 2019). GSTP1 substrate consists of reactive metabolites derived from ifosfamide, busulfan and chlorambucil (Allan et al. 2001). The various properties of this isoenzyme include electrophilic compound detoxification, chaperone function, regulation of nitric oxide and kinase signaling pathways, as well as protein S-glutathionylation (López-Rodríguez et al. 2019). It is known that GSTP can bind to different proteins, such as NF-kB, JNK and p53, and is involved in the regulation of inflammation, cell proliferation and apoptosis (López-Rodríguez et al. 2019). Furthermore, this enzyme plays a crucial role in the release of leukotrienes and prostaglandins (Oh et al. 2005). It has been shown that the exposure of lung epithelial cells to lipopolysaccharide induces NF-kB activation mediated by GSTP (López-Rodríguez et al. 2019).

The different *GSTP1* alleles, *GSTP1* A-D, result from the *GSTP1* Ile105Val and Ala114Val polymorphisms (López-Rodríguez et al. 2019). These substitutions cause variations in substrate-specific activity and thermal stability in the encoded proteins (Gsur et al. 2001). The Ile105Val substitution at the substrate binding-site of GSTP1 leads to an alteration in substrate-specific catalytic activity and thermal instability (Allan et al. 2001). This Val allele variant, related to reduced enzyme activity (Safarinejad, Shafiei, and Safarinejad 2011), has also been associated with a higher risk of allergic asthma, as well as lung dysfunction (Oh et al. 2005; López-Rodríguez et al. 2019). Moreover, *GST* polymorphisms related to diminished or absent enzyme activity hinder the detoxification of toxic compounds in the airways, increasing airway damage and affecting lung development (Kabesch et al. 2004).

Recently, López-Rodríguez and colleagues demonstrated the secretion of GSTP by bronchial epithelial cells, observing that GSTP enhances the cysteine-protease activity of Der p 1, an allergen present in house dust mites (López-Rodríguez et al. 2019). This suggests the clinical implication of GSTP, since individuals with high GSTP activity in the airways, in the presence of Der p 1, may present sensitization and an exacerbated immune response. Moreover, elevated levels of GSTP were also observed in the plasma as well as biopsies of patients with esophageal and gastric cancer, and also in the bronchoalveolar lavage of patients with lung disease (López-Rodríguez et al. 2019). Reports have demonstrated that the *GSTP1* Ile105Val polymorphism is associated with asthma development, as well as asthma-related phenotypes, including atopy and airway hyperresponsiveness to drugs such as methacholine (Oh et al. 2005). On the contrary, other studies found that the *GSTP1* Ile105 allele plays a protective role in asthma development (Oh et al. 2005). A study performed in the Korean population found no associations between the *GSTP1* Ile105Val polymorphism and aspirin-intolerant asthma (Oh et al. 2005), defined as the development of asthma due to intake of aspirin or anti-inflammatory drugs (López-Rodríguez et al. 2019).

Oh and colleagues reported that the *GSTP1* Ile105Ile genotype might be associated with good prognosis in patients with rheumatoid arthritis and basal cell carcinoma (Oh et al. 2005). CpG island hypermethylation in *GSTP*, related to the lack of GSTP1 enzyme, has also been associated with prostate cancer (Safarinejad, Shafiei, and Safarinejad 2011). Studies performed in men demonstrated an association between the *GSTP1* 105Val allele and lower prostate cancer risk (Gsur et al. 2001; Sivoňová et al. 2009).

GST polymorphisms have been linked to acute myeloid leukemia susceptibility. Allan and colleagues demonstrated that the *GSTP1* 105Val allele variant was associated with a higher risk of developing therapy-related acute myeloid leukemia (t-AML), specifically after chemotherapy involving derivatives of cyclophosphamide, chlorambucil, doxorubicin, etoposide and cisplatin (Allan et al. 2001). Indeed, these authors found no association between the variant *GSTP1* 105Val allele and t-AML in patients submitted to radiotherapy. It is important to note that these chemotherapeutics are present in the treatment regimens of several types of cancer, including lymphatic, bladder, breast, ovarian, lung and testicular (Allan et al. 2001). Accordingly, the *GSTP1* Ile105Val polymorphism may be considered as a biomarker of t-AML after chemotherapy and could aid in the early and intensive surveillance of patients undergoing therapy with these drugs (Allan et al. 2001).

GSTs and Treatment Response

In addition to protecting cells from environmental and oxidative stress, GSTs are employed by malign cells to escape the cytotoxic effects of anticancer drugs (McIlwain, Townsend, and Tew 2006). It has been suggested that chemotherapeutic resistance in cancer cells can occur due to different mechanisms, including the detoxification of toxic metabolites (Martos-Maldonado et al. 2012). Safarinejad and colleagues suggested that high GST activity, i.e., ultra-rapid drug metabolism could reduce the bioavailability of intermediate metabolites, and consequently the cytotoxic effects of anticancer drugs on tumor cells (Safarinejad, Shafiei, and Safarinejad 2011). Moreover, these authors found that *GST* genotypes linked to low or no GST activity were associated with cancer survival.

hGSTA1 has been associated with cellular resistance to antineoplastic drugs, insecticides, herbicides and antibiotics (Martínez-Guzmán et al. 2017). It was suggested that the resistance of cancer cells to chemotherapeutic agents could be linked to higher GSTP1-1 activity, since this isoenzyme is the GST most expressed by these cells (Martos-Maldonado et al. 2012). *GSTT1* polymorphisms have also been found to be associated with chemotherapy nonresponse (Zhou et al. 2014), as previous reports demonstrated an association between *GSTM1* and *T1* polymorphisms and treatment outcome in patients with breast cancer, as well as children with leukemia (Ambrosone et al. 2001; Davies et al. 2001). However, in patients with colorectal cancer, this association was not found (Stoehlmacher et al. 2002). Abbas and colleagues demonstrated that cervical cancer patients, who were carriers of the *GSTM1* and *T1* null genotypes as well as the *GSTP1-1* Val allele variant, presented better overall survival in response to cisplatin-based concomitant chemoradiation, suggesting that the detection of these polymorphisms may be used as prognostic markers in these patients (Abbas et al. 2015). Corroborating this finding, other studies have also linked these polymorphisms to treatment outcomes in different cancers treated with platinum agents and radiation. In 2009, Singh and colleagues observed a significant differences in blood levels of CsA in patients with the *GSTM1+* and *GSTM1* null genotypes

(Singh et al. 2009). Patients who were carriers of the *GSTM1* null genotype that were submitted to renal transplantation and treated with cyclosporine presented a higher risk of rejection compared to those treated with tacrolimus, suggesting that patients with this genotype require lower doses of cyclosporine (Akgul et al. 2012).

GSTP1-1 may inhibit apoptosis by binding to c-Jun N-terminal kinase (JNK), a member of the mitogen-activated protein kinase family, and the scaffold protein TNF-α-receptor-associated-factor 2 (TRAF2) (Sau et al. 2010). It follows that the inhibition of this isoenzyme may represent a promising strategy (Allocati et al. 2018). Building on this hypothesis, Di Paolo and colleagues synthesized and characterized a new benzamide-containing nitrobenzoxadiazole that inhibits GSTP1-1 and presents higher stability to human liver microsomal carboxylesterases than the parent compound, an inhibitor of GSTP1-1. Experimental studies showed that this new molecule presented a greater cytotoxic effect on melanoma cells (Di Paolo et al. 2019). Another study demonstrated the importance of GSTP1 in cellular resistance to metabolites, such as platinum derivatives, etoposide, cyclophosphamide, melphalan and doxorubicin (Allan et al. 2001).

Similarly to cancer cells, the secretion of antioxidant and detoxification enzymes, such as GSTs, aids in nematodes' evasion of host immune response. Experiments performed in nematodes at different stages revealed that the inhibition of GST was followed by a significant reduction in parasite viability (Rajaiah Prabhu et al. 2018). Some investigators have suggested that the GST produced by *Wuchereria bancrofti* might be a target for therapeutic agents, due to its crucial role in the detoxification of toxic metabolites (Rajaiah Prabhu et al. 2018). *Plasmodium falciparum* cytosolic GST (*Pf*GST) is a crucial enzyme for the detoxification of toxic heme (Al-Qattan, Mordi, and Mansor 2016). An experimental study demonstrated the selective inhibition of parasite GST by antimalarial agents and suggested that this strategy may prove beneficial in malaria chemotherapy (Chikezie 2011). Indeed, the inhibition of parasite GST activity leads to the accumulation of antimalarial metabolites within parasites, creating a hostile milieu that culminates in parasite death.

Furthermore, higher levels of *Pf*GST were observed in parasites with demonstrated resistance to chloroquine (Al-Qattan, Mordi, and Mansor 2016). These findings suggest that GSTs might be considered as potential targets for anthelmintic, antiparasitic and anticancer drug discovery (Al-Qattan, Mordi, and Mansor 2016).

CONCLUSION

GSTs are phase II enzymes that catalyze the biotransformation of a wide range of electrophilic compounds. These enzymes play a crucial role in the detoxification of toxic metabolites produced during phase I reactions. The characterization of human and parasite GSTs will be extremely important to the discovery and/or design of antineoplastic and antiparasitic agents. Moreover, for patients undergoing treatment, individualized therapy made possible through the investigation of the effects caused by genetic alterations in *GST* genes, as well as genetic polymorphisms in genes encoding phase I enzymes responsible for the production of intermediate metabolites, will be of great benefit. The thorough elucidation of these aspects of GST should also reduce the incidence of drug resistance observed in patients with cancer, and also that presented by parasites in the host. Although some studies have identified some *GST* polymorphisms as biomarkers of certain pathologies, such as hepatotoxicity, nephropathy, neuropathy, asthma, cancer and its recurrence, as well as drug resistance, the specific relationships between many *GST* polymorphisms and related pathogeneses nonetheless requires validation.

REFERENCES

Abbas, Mohammad, Vandana Singh Kushwaha, Kirti Srivastava, and Monisha Banerjee. 2015. "Glutathione S-Transferase Gene Polymorphisms and Treatment Outcome in Cervical Cancer Patients under Concomitant Chemoradiation." Edited by Ken Mills. *PLOS ONE* 10 (11): e0142501. https://doi.org/10.1371/journal.pone.0142501.

Ahn, J., Marilie D.Gammon, Regina M.Santella, Mia M.Gaudet, Julie A.Britton, Susan L.Teitelbaum, Mary Beth Terry, et al. 2006. "Effects of Glutathione S-Transferase A1 (GSTA1) Genotype and Potential Modifiers on Breast Cancer Risk." *Carcinogenesis* 27 (9): 1876–82. https://doi.org/10.1093/carcin/bgl038.

Akgul, S.U., F.S. Oguz, Y. Çalişkan, C. Kekik, H. Gürkan, A. Türkmen, I. Nane, and F. Aydin. 2012. "The Effect of Glutathion S-Transferase Polymoprhisms and Anti-GSST1 Antibodies on Allograft Functions in Recipients of Renal Transplant." *Transplantation Proceedings* 44 (6): 1679–84. https://doi.org/10.1016/j.transproceed.2012.04.004.

Allan, J. M., C. P. Wild, S. Rollinson, E. V. Willett, A. V. Moorman, G. J. Dovey, P. L. Roddam, E. Roman, R. A. Cartwright, and G. J. Morgan. 2001. "Polymorphism in Glutathione S-Transferase P1 Is Associated with Susceptibility to Chemotherapy-Induced Leukemia." *Proceedings of the National Academy of Sciences* 98 (20): 11592–97. https://doi.org/10.1073/pnas.191211198.

Allocati, Nerino, Michele Masulli, Carmine Di Ilio, and Luca Federici. 2018. "Glutathione Transferases: Substrates, Inihibitors and pro-Drugs in Cancer and Neurodegenerative Diseases." *Oncogenesis* 7 (1): 8. https://doi.org/10.1038/s41389-017-0025-3.

Al-Qattan, Mohammed Nooraldeen, Mohd Nizam Mordi, and Sharif Mahsofi Mansor. 2016. "Assembly of Ligands Interaction Models for Glutathione-S-Transferases from Plasmodium Falciparum, Human and Mouse Using Enzyme Kinetics and Molecular Docking." *Computational Biology and Chemistry* 64 (October): 237–49. https://doi.org/10.1016/j.compbiolchem.2016.07.007.

Ambrosone, C. B., C. Sweeney, B. F. Coles, P. A. Thompson, G. Y. McClure, S. Korourian, M. Y. Fares, A. Stone, F. F. Kadlubar, and L. F. Hutchins. 2001. "Polymorphisms in Glutathione S-Transferases (GSTM1 and GSTT1) and Survival after Treatment for Breast Cancer." *Cancer Research* 61 (19): 7130–35.

Barjui, Shahrbanou Parchami, Somayeh Reiisi, and Asghar Bayati. 2017. "Human Glutathione S-Transferase Enzyme Gene Variations and Risk of Multiple Sclerosis in Iranian Population Cohort." *Multiple Sclerosis and Related Disorders* 17 (October): 41–46. https://doi.org/10.1016/j.msard.2017.06.016.

Bock, Karl Walter. 2014. "Homeostatic Control of Xeno- and Endobiotics in the Drug-Metabolizing Enzyme System." *Biochemical Pharmacology* 90 (1): 1–6. https://doi.org/10.1016/j.bcp.2014.04.009.

Božina, Nada, Vlasta Bradamante, and Mila Lovrić. 2009. "Genetic Polymorphism of Metabolic Enzymes P450 (CYP) as a Susceptibility Factor for Drug Response, Toxicity, and Cancer Risk." *Archives of Industrial Hygiene and Toxicology* 60 (2). https://doi.org/10.2478/10004-1254-60-2009-1885.

Braver, Michiel W. den, Yongjie Zhang, Harini Venkataraman, Nico P.E. Vermeulen, and Jan N.M. Commandeur. 2016. "Simulation of Interindividual Differences in Inactivation of Reactive Para - Benzoquinone Imine Metabolites of Diclofenac by Glutathione S - Transferases in Human Liver Cytosol." *Toxicology Letters* 255 (July): 52–62. https://doi.org/10.1016/j.toxlet.2016.05.015.

Chen, Rui, Shancheng Ren, Tong Meng, Josephine Aguilar, and Yinghao Sun. 2013. "Impact of Glutathione-S-Transferases (GST) Polymorphisms and Hypermethylation of Relevant Genes on Risk of Prostate Cancer Biochemical Recurrence: A Meta-Analysis." Edited by Olga Y. Gorlova. *PLoS ONE* 8 (9): e74775. https://doi.org/10.1371/journal.pone.0074775.

Chikezie, Paul Chidoka. 2011. "Glutathione S-Transferase Activity of Human Erythrocytes Incubated in Aqueous Solutions of Five Antimalarial Drugs." *Free Radicals and Antioxidants* 1 (2): 26–30. https://doi.org/10.5530/ax.2011.2.6.

Coles, Brian F., and Fred F. Kadlubar. 2005. "Human Alpha Class Glutathione S-Transferases: Genetic Polymorphism, Expression, and Susceptibility to Disease." In *Methods in Enzymology*, 401:9–42. Elsevier. https://doi.org/10.1016/S0076-6879(05)01002-5.

Cook, D. G, and D. P Strachan. 1999. "Health Effects of Passive Smoking 10: Summary of Effects of Parental Smoking on the Respiratory Health of Children and Implications for Research." *Thorax* 54 (4): 357–66. https://doi.org/10.1136/thx.54.4.357.

Davies, Stella M., Leslie L. Robison, Jonathan D. Buckley, Tom Tjoa, William G. Woods, Gretchen A. Radloff, Julie A. Ross, and John P. Perentesis. 2001. "Glutathione S-Transferase Polymorphisms and Outcome of Chemotherapy in Childhood Acute Myeloid Leukemia." *Journal of Clinical Oncology* 19 (5): 1279–87. https://doi.org/10.1200/JCO.2001.19.5.1279.

Di Paolo, Veronica, Chiara Fulci, Dante Rotili, Francesca Sciarretta, Alessia Lucidi, Blasco Morozzo della Rocca, Anastasia De Luca, Antonio Rosato, Luigi Quintieri, and Anna Maria Caccuri. 2019. "Synthesis and Characterisation of a New Benzamide-Containing Nitrobenzoxadiazole as a GSTP1-1 Inhibitor Endowed with High Stability to Metabolic Hydrolysis." *Journal of Enzyme Inhibition and Medicinal Chemistry* 34 (1): 1131–39. https://doi.org/10.1080/14756366.2019.1617287.

Gsur, Andrea, Gerald Haidinger, Sonja Hinteregger, Gabriele Bernhofer, Georg Schatzl, Stephan Madersbacher, Michael Marberger, Christian Vutuc, and Michael Micksche. 2001. "Polymorphisms of Glutathione-S-transferase Genes (GSTP1, GSTM1 and GSTT1) and Prostate-cancer Risk." *International Journal of Cancer* 95: 152–55.

Gulçin, İlhami, Parham Taslimi, Ayşenur Aygün, Nastaran Sadeghian, Enes Bastem, Omer Irfan Kufrevioglu, Fikret Turkan, and Fatih Şen. 2018. "Antidiabetic and Antiparasitic Potentials: Inhibition Effects of Some Natural Antioxidant Compounds on α-Glycosidase, α-Amylase and Human Glutathione S-Transferase Enzymes." *International Journal of Biological Macromolecules* 119 (November): 741–46. https://doi.org/10.1016/j.ijbiomac.2018.08.001.

Jansson, Mattias, Alvaro Rada, Lidija Tomic, Lill-Inger Larsson, and Claes Wadelius. 2003. "Analysis of the Glutathione S-Transferase M1 Gene Using Pyrosequencing and Multiplex PCR–No Evidence of Association to Glaucoma." *Experimental Eye Research* 77 (2): 239–43. https://doi.org/10.1016/S0014-4835(03)00109-X.

Kabesch, M, C Hoefler, D Carr, W Leupold, S K Weiland, and E von Mutius. 2004. "Glutathione S Transferase Deficiency and Passive Smoking Increase Childhood Asthma." *Thorax* 59 (7): 569–73. https://doi.org/10.1136/thx.2003.016667.

Koumaravelou, K., Z. Shoaib, C. Adithan, D. Charron, A. Srivastava, R. Tamouza, and R. Krishnamoorthy. 2011. "Analysis of Human Glutathione S-Transferase Alpha 1 (HGSTA1) Gene Promoter Polymorphism Using Denaturing High Performance Liquid Chromatography (DHPLC)." *Clinica Chimica Acta* 412 (15–16): 1465–68. https://doi.org/10.1016/j.cca.2011.04.019.

Lawless, Matthew J., John R. Pettersson, Gordon S. Rule, Frederick Lanni, and Sunil Saxena. 2018. "ESR Resolves the C Terminus Structure of the Ligand-Free Human Glutathione S-Transferase A1-1." *Biophysical Journal* 114 (3): 592–601. https://doi.org/10.1016/j.bpj.2017.12.016.

López-Rodríguez, Juan Carlos, Juliana Manosalva, J. Daniel Cabrera-García, María M. Escribese, Mayte Villalba, Domingo Barber, Antonio Martínez-Ruiz, and Eva Batanero. 2019. "Human Glutathione-S-Transferase Pi Potentiates the Cysteine-Protease Activity of the Der p 1 Allergen from House Dust Mite through a Cysteine Redox Mechanism." *Redox Biology* 26 (September): 101256. https://doi.org/10.1016/j.redox.2019.101256.

Marcus, Carol J., William H. Habig, and William B. Jakoby. 1978. "Glutathione Transferase from Human Erythrocytes." *Archives of Biochemistry and Biophysics* 188 (2): 287–93. https://doi.org/10.1016/S0003-9861(78)80011-3.

Martínez-Guzmán, Carmen, Pedro Cortés-Reynosa, Eduardo Pérez-Salazar, and Guillermo Elizondo. 2017. "Dexamethasone Induces Human Glutathione S Transferase Alpha 1 (HGSTA1) Expression through the Activation of Glucocorticoid Receptor (HGR)." *Toxicology* 385 (June): 59–66. https://doi.org/10.1016/j.tox.2017.05.002.

Martos-Maldonado, Manuel C., Indalecio Quesada-Soriano, Federico García-Maroto, Antonio Vargas-Berenguel, and Luís García-Fuentes. 2012. "Ferrocene Labelings as Inhibitors and Dual Electrochemical Sensors of Human Glutathione S-Transferase P1-1." *Bioorganic & Medicinal Chemistry Letters* 22 (23): 7256–60. https://doi.org/10.1016/j.bmcl.2012.09.022.

McIlwain, C C, D M Townsend, and K D Tew. 2006. "Glutathione S-Transferase Polymorphisms: Cancer Incidence and Therapy." *Oncogene* 25 (11): 1639–48. https://doi.org/10.1038/sj.onc.1209373.

Nguyen, Thai V., Marcel J. R. Janssen, Martijn G. H. van Oijen, Saskia M. Bergevoet, Rene H. M. te Morsche, Henri van Asten, Robert J. F. Laheij, Wilbert H. M. Peters, and Jan B. M. J. Jansen. 2009. "Genetic Polymorphisms in GSTA1, GSTP1, GSTT1, and GSTM1 and Gastric Cancer Risk in a Vietnamese Population." *Oncology Research Featuring Preclinical and Clinical Cancer Therapeutics* 18 (7): 349–55. https://doi.org/10.3727/096504010X12626118080064.

Nock, Nora L., Cathryn Bock, Christine Neslund-Dudas, Jennifer Beebe-Dimmer, Andrew Rundle, Deliang Tang, Michelle Jankowski, and Benjamin A. Rybicki. 2009. "Polymorphisms in Glutathione S-Transferase Genes Increase Risk of Prostate Cancer Biochemical Recurrence Differentially by Ethnicity and Disease Severity." *Cancer Causes & Control* 20 (10): 1915–26. https://doi.org/10.1007/s10552-009-9385-0.

Oh, Jung-Mi, Seung-Hyun Kim, Chang-Hee Suh, Dong-Ho Nahm, Hae-Sim Park, Young-Mok Lee, June-Hyuk Lee, Choon-Sik Park, and Hyung-Doo Shin. 2005. "Lack of Association of Glutathione S-Transferase P1 Ile105Val Polymorphism with Aspirin-Intolerant Asthma." *The Korean Journal of Internal Medicine* 20 (3): 232. https://doi.org/10.3904/kjim.2005.20.3.232.

Pagliuso, R.G., M. Abbud-Filho, M.P.S. Alvarenga, M.A.S. Ferreira-Baptista, J.M. Biselli, P.M. Biselli, E.M. Goloni-Bertollo, and E.C. Pavarino-Bertelli. 2008. "Role of Glutathione S-Transferase Polymorphisms and Chronic Allograft Dysfunction." *Transplantation Proceedings* 40 (3): 743–45. https://doi.org/10.1016/j.transproceed.2008.03.008.

Poppel, Geert van, Nico de Vogel, Peter J. van Balderen, and Frans J. Kok. 1992. "Increased Cytogenetic Damage in Smokers Deficient in Glutathione S-Transferase Isozyme μ." *Carcinogenesis* 13 (2): 303–5. https://doi.org/10.1093/carcin/13.2.303.

Rajaiah Prabhu, Prince, Sakthi Devi Moorthy, Jayaprakasam Madhumathi, Satya Narayan Pradhan, Markus Perbandt, Christian Betzel, and Perumal Kaliraj. 2018. "Wucherria Bancrofti Glutathione S-Transferase: Insights into the 2.3 Å Resolution X-Ray Structure and Function, a Therapeutic Target for Human Lymphatic Filariasis."

Biochemical and Biophysical Research Communications 505 (4): 979–84. https://doi.org/10.1016/j.bbrc.2018.09.077.

Ricci, G., G. Del Boccio, A. Pennelli, A. Aceto, E. P. Whitehead, and G. Federici. 1989. "Nonequivalence of the Two Subunits of Horse Erythrocyte Glutathione Transferase in Their Reaction with Sulfhydryl Reagents." *The Journal of Biological Chemistry* 264 (10): 5462–67.

Ryberg, D, V Skaug, A Hewer, DH Phillips, LW Harries, CR Wolf, D Ogreid, A Ulvik, P Vu, and A Haugen. 1997. "Genotypes of Glutathione Transferase M1 and P1 and Their Significance for Lung DNA Adduct Levels and Cancer Risk." *Carcinogenesis* 18 (7): 1285–89. https://doi.org/10.1093/carcin/18.7.1285.

Sá, Renata Almeida de, Aline dos Santos Moreira, Pedro Hernan Cabello, Antonio Augusto Ornellas, Eduardo Butinhão Costa, Cintia da Silva Matos, Gilda Alves, and Ana Hatagima. 2014. "Human Glutathione S-Transferase Polymorphisms Associated with Prostate Cancer in the Brazilian Population." *International Braz j Urol* 40 (4): 463–73. https://doi.org/10.1590/S1677-5538.IBJU.2014.04.04.

Safarinejad, M R, N Shafiei, and S H Safarinejad. 2011. "Glutathione S-Transferase Gene Polymorphisms (GSTM1, GSTT1, GSTP1) and Prostate Cancer: A Case-Control Study in Tehran, Iran." *Prostate Cancer and Prostatic Diseases* 14 (2): 105–13. https://doi.org/10.1038/pcan.2010.54.

Salcedo, Magdalena, Margarita Rodríguez-Mahou, Carmen Rodríguez-Sainz, Diego Rincón, Emilio Alvarez, Jose Luis Vicario, Maria-Vega Catalina, et al. 2009. "Risk Factors for Developing de Novo Autoimmune Hepatitis Associated with Anti-Glutathione S-Transferase T1 Antibodies after Liver Transplantation." *Liver Transplantation* 15 (5): 530–39. https://doi.org/10.1002/lt.21721.

Sau, Andrea, Francesca Pellizzari Tregno, Francesco Valentino, Giorgio Federici, and Anna Maria Caccuri. 2010. "Glutathione Transferases and Development of New Principles to Overcome Drug Resistance." *Archives of Biochemistry and Biophysics* 500 (2): 116–22. https://doi.org/10.1016/j.abb.2010.05.012.

Sergentanis, Theodoros N., and Konstantinos P. Economopoulos. 2010. "GSTT1 and GSTP1 Polymorphisms and Breast Cancer Risk: A Meta-

Analysis." *Breast Cancer Research and Treatment* 121 (1): 195–202. https://doi.org/10.1007/s10549-009-0520-0.

Singh, Ranjana, Parmeet K. Manchanda, Pravin Kesarwani, Aneesh Srivastava, and Rama D. Mittal. 2009. "Influence of Genetic Polymorphisms in GSTM1, GSTM3, GSTT1 and GSTP1 on Allograft Outcome in Renal Transplant Recipients." *Clinical Transplantation* 23 (4): 490–98. https://doi.org/10.1111/j.1399-0012.2009.00985.x.

Sivoňová, Monika, Iveta Waczulíková, Dušan Dobrota, Tatiana Matáková, Jozef Hatok, Peter Račay, and Ján Kliment. 2009. "Polymorphisms of Glutathione-S-Transferase M1, T1, P1 and the Risk of Prostate Cancer: A Case-Control Study." *Journal of Experimental & Clinical Cancer Research* 28 (1): 32. https://doi.org/10.1186/1756-9966-28-32.

Srivastava, Daya Shankar Lal, Anil Mandhani, Balraj Mittal, and Rama Devi Mittal. 2005. "Genetic Polymorphism of Glutathione S-Transferase Genes (GSTM1, GSTT1 and GSTP1) and Susceptibility to Prostate Cancer in Northern India." *BJU International* 95 (1): 170–73. https://doi.org/10.1111/j.1464-410X.2005.05271.x.

Stanulla, Martin, Martin Schrappe, Annette Müller Brechlin, Martin Zimmermann, and Karl Welte. 2000. "Polymorphisms within Glutathione S-Transferase Genes (GSTM1, GSTT1, GSTP1) and Risk of Relapse in Childhood B-Cell Precursor Acute Lymphoblastic Leukemia: A Case-Control Study." *Blood* 95 (4): 1222–28. https://doi.org/10.1182/blood.V95.4.1222.004k20_1222_1228.

Stoehlmacher, J., David J. Park, Wu Zhang, Susan Groshen, Denice D. Tsao-Wei, Mimi C. Yu, and Heinz-Josef Lenz. 2002. "Association Between Glutathione S-Transferase P1, T1, and M1 Genetic Polymorphism and Survival of Patients With Metastatic Colorectal Cancer." *Journal of the National Cancer Institute* 94 (12): 936–42. https://doi.org/10.1093/jnci/94.12.936.

Sundberg, Kathrin, Kristian Dreij, Albrecht Seidel, and Bengt Jernström. 2002. "Glutathione Conjugation and DNA Adduct Formation of Dibenzo[a,1]Pyrene and Benzo[a]Pyrene Diol Epoxides in V79 Cells Stably Expressing Different Human Glutathione Transferases." *Chemical Research in Toxicology* 15 (2): 170–79. https://doi.org/10.1021/tx015546t.

Tourinho-Barbosa, Rafael, Victor Srougi, Igor Nunes-Silva, Mohammed Baghdadi, Gregory Rembeyo, Sophie S. Eiffel, Eric Barret, et al. 2018. "Biochemical Recurrence after Radical Prostatectomy: What Does It Mean?" *International Braz j Urol* 44 (1): 14–21. https://doi.org/10.1590/s1677-5538.ibju.2016.0656.

Wichmann, Ingeborg, Isabel Aguilera, Jose M. Sousa, Angel Bernardos, Emilio J. Garcia Nunez‡, Eduardo Vigil, Rosario Magarino, Isabel Magarino, Antonia Torres, and Antonio Nunez-Roldan. 2006. "Antibodies against Glutathione S-Transferase T1 in Non?Solid Organ Transplanted Patients." *Transfusion* 46 (9): 1505–9. https://doi.org/10.1111/j.1537-2995.2006.00938.x.

Yang, J., G. Tezel, R. V. Patil, C. Romano, and M. B. Wax. 2001. "Serum Autoantibody against Glutathione S-Transferase in Patients with Glaucoma." *Investigative Ophthalmology & Visual Science* 42 (6): 1273–76.

Zhang, Yongjie, Shalenie P. den Braver-Sewradj, J. Chris Vos, Nico P.E. Vermeulen, and Jan N.M. Commandeur. 2017. "Human Glutathione S-Transferases- and NAD(P)H:Quinone Oxidoreductase 1-Catalyzed Inactivation of Reactive Quinoneimines of Amodiaquine and N-Desethylamodiaquine: Possible Implications for Susceptibility to Amodiaquine-Induced Liver Toxicity." *Toxicology Letters* 275 (June): 83–91. https://doi.org/10.1016/j.toxlet.2017.05.003.

Zhou, T.-B., G. P. C. Drummen, Z.-P. Jiang, and Y.-H. Qin. 2014. "GSTT1 Polymorphism and the Risk of Developing Prostate Cancer." *American Journal of Epidemiology* 180 (1): 1–10. https://doi.org/10.1093/aje/kwu112.

In: Glutathione S-Transferases
Editor: Igor Azevedo Silva

ISBN: 978-1-53618-188-3
© 2020 Nova Science Publishers, Inc.

Chapter 2

GENETICS OF GLUTATHIONE S-TRANSFERASES AND COMPLEX DISEASES

Atar Singh Kushwah[1,2,*] *and Monisha Banerjee*[1,†]

[1]Molecular and Human Genetics Laboratory, Department of Zoology, University of Lucknow, Lucknow, India
[2]Department of Zoology, Institute of Science, Banaras Hindu University, Varanasi, India

ABSTRACT

The glutathione-S-transferases (GSTs) are dimeric cytosolic xenobiotic-metabolizing enzymes that catalyze the conjugation of an active xenobiotic to GSH, an endogenous water-soluble substrate. The GST-mediated detoxification pathway ensures cellular protection from environmental insults and oxidative stress. Cellular ROS/RNS levels promote peroxidation of lipids and lipoproteins present in biomembranes, which leads to development of several pathological conditions. GSTs protect cells against deleterious actions of ROS/ RNS by promoting redox

---

[*] Corresponding Author's E-mail: as_kushwah@hotmail.com.
[†] Corresponding Author's E-mail: monishabanerjee30@gmail.com.

homeostasis through neutralization of excessive reactive metabolites, whose chemical actions elicit numerous signaling cascades associated with cell proliferation, inflammatory responses, apoptosis and senescence. Human GSTs are divided into three main families, namely, cytosolic (α-, β-, γ-, δ-, ξ-and ζ), mitochondrial (π) and membrane-bound (κ).These classes of GSTs originate from different chromosomes but share ~30% sequence identity and have cell specific distribution.

Genetic variants in antioxidant enzyme encoding genes lead to decreased enzymatic activity leading to oxidative DNA damage and chronic inflammation. Moreover, genetic variants in *GSTM1, T1* and *P1* have been reported to be involved in development of type 2 diabetes mellitus (T2DM) and various cancers like cervical (CaCx), bladder, colon, gastrointestinal, lung, pulmonary diseases, neurological disorders and various pathological phenotypes. Interestingly, some of the common genetic variants like *GSTM1*-Del (16kb deletion), *GSTT1*-Del (54kbdeletion) polymorphisms (Null/Present) and *GSTP1* +313A/G (105I/V rs1695) are significantly associated with complex diseases *viz.* T2DM and CaCx in north Indian population as well as other ethnic groups. The *GST* genetic variants have also shown interesting association with treatment outcome in cervical cancer patients undergoing chemoradiotherapy *viz.* clinical response, vital status, overall survival and toxicity.

In addition to the above, this chapter focuses on the importance of functional aspect of polymorphic variants in different populations so that the risk genotypes can be identified as indicators of disease susceptibility. Genetic studies will help to develop prognostic biomarkers for early prediction and risk assessment in patients and enable clinicians to develop personalized treatment regimes.

Keywords: cervical cancer, clinical application, disease susceptibility, genetic variants, glutathione-S-transferases, type 2 diabetes mellitus

BACKGROUND

The glutathione-S-transferases (GSTs) are xenobiotic-metabolizing enzymes that catalyze the conjugation reaction (xenobiotic to GSH). They detoxify reactive electrophiles such as those contained in tobacco smoke. The GST-mediated detoxification pathways protect cells from environmental insults and oxidative stress. It has also been implicated in cellular resistance to drugs (Abbas et al., 2013; 2015).

Reactive Oxygen Species (ROS)/Reactive Nitrogen Species (RNS) levels promote peroxidation of lipids and lipoproteins, which leads to development of several pathological conditions (Nebert and Timothy, 2006). The GSTs protect the cell against deleterious actions of ROS/RNS by promoting redox homeostasis through neutralization of excessive reactive metabolites, whose chemical actions elicit numerous signaling cascades associated with cell proliferation, inflammatory responses, apoptosis and senescence (Banerjee and Vats, 2014).

In literature, microsomal GSTs (Morgenstern et al., 1983) and leukotriene C4 synthase (Nicholson et al., 1993) are also described as members of the GST family although they do not show any significant sequence homology with cytosolic GSTs (Hayes, et al., 1995). GSTs are widely distributed in nature from unicellular to multicellular organisms *viz* bacteria, yeast, plants and animals (Sheehan et al., 2001). In plants, GSTP (phi), GST Tau, GSTT (theta), GSTZ (zeta), GST lambda, GSTT and GSTZ have counterparts in animals (Dixon et al., 2002). GST sigma and GSTT classes are more abundant in non-vertebrate animals (Hayes, et al., 1995). La Roche et al. (1990), suggesting that the progenitor of mammalian GSTs comes from GSTT class which have significant homology with *Dichloromethane dehalogenase* enzyme from the Methylobacterium (La Roche et al., 1990).

Human GSTs are divided into three main families, namely, cytosolic GSTs (α-, β-, γ-, δ-, ξ-and ζ), mitochondrial GST (π) and membrane-bound GST (κ) (Kamisaka et al., 1975). These classes of GSTs originate from different chromosomes but share ~30% sequence identity and have cell specific distribution (Board et al., 1997). The cytosolic hGSTP1 represents 95% of its GST pool, which is a homodimeric intracellular protein (46 kDa) expressed in different organs in a cell specific manner (Tew et al., 2011). In general, the cytosolic GST monomers range from 22-29 KDa and have a variety of substrates. Each monomer of GST has an active site which contains two sub-sites; one is the less conserved H site for hydrophobic substrates and second is the highly conserved G site for GSH binding (Rushmore et al., 1993).

FUNCTIONS OF GLUTATHIONE *S*-TRANSFERASES

GSTs are dimeric (45-55KDa) isoenzymes, cytosolic enzymes which have been assigned as α-, μ-, π-, θ-, κ- Ω- and δ-GSTs. They have extensive ligand binding properties in addition to their catalytic role in detoxification (Listowsky et al., 1988; Sheehan et al., 2001). They have also been implicated in a variety of resistance phenomena involving cancer chemotherapy agents (McLellan et al., 1999), insecticides (Ranson et al., 1997), herbicides (Edwards et al., 2000) and microbial antibiotics (Arca et al., 1997). The GSTs comprise a complex and widespread enzyme super family that has been subdivided further into an ever-increasing number of classes based on a variety of criteria, including amino acid/nucleotide sequence, immunological, kinetic and tertiary/quaternary structural properties (Nebert and Timothy, 2006). In addition to their catalytic role in detoxification, GSTs were also found to possess selenium independent peroxidase activity with hydroperoxides, steroid isomerization capacity, binding and transport of bilirubin, heme, bile salts and steroids in a process that is associated with a decrease in enzymatic activity (Sheehan et al., 2001) (Table 1).

GSTs are the phase II detoxification enzymes which catalyze the conjugation of electrophilic substrates to glutathione (GSH) and have peroxidase, isomerase activities which protect cells against H_2O_2-induced cell death by inhibiting the Jun N-terminal kinase (Banerjee and Vats, 2014). They are also able to bind non-catalytically to a wide range of endogenous and exogenous ligands. Cytosolic GSTs of Alpha, Mu, Pi and Theta classes are well characterized on the basis of their combination of substrate/inhibitor specificity, primary and tertiary structure similarities and immunological identities (Table 1).

Table 1. Glutathione S-transferase encoding genes, chromosomal/genomic location, gene size, transcription factors and functions

S. No.	Gene	Chromosomal/ Genomic location	Gene size (in bases)	Transcription Factors	Function
1	*GSTA1* (Glutathione S-transferase alpha 1)	6p12.2 52791371-52803866	12,496	AP-1, AP-4, ATF-2, c-Jun, FOXD1, HNF-1, HNF-1A, Lmo2, Nkx2-5, Sp1	Addition of glutathione to target electrophilic compounds, including carcinogens, therapeutic drugs, environmental toxins, and products of oxidative stress. Role in protecting cells from reactive oxygen species and the products of peroxidation.
2	*GSTA2* (Glutathione S-transferase alpha 2)	6p12.2 52750087-52763563	13,477	C/EBPalpha,FOXD1, FOXD3,HFH-1,HNF-1,HNF-1A,HNF-3β,Pbx1a,PPAR-1,PPAR-gamma2	Conjugation of reduced glutathione to a wide number of exogenous and endogenous hydrophobic electrophiles.
3	*GSTA3* (Glutathione S-Transferase Alpha 3)	6p12.2 52896639-52909798	13,160	En-1 Ik-2 STAT3 YY1	Catalyzes isomerization reactions that contribute to the biosynthesis of steroid hormones. Efficiently catalyze obligatory double-bond isomerization of delta(5)-androstene-3,17-dione and delta(5)-pregnene-3,20-dione, precursors to testosterone and progesterone

Table 1. (Continued)

S. No.	Gene	Chromosomal/Genomic location	Gene size (in bases)	Transcription Factors	Function
4	GSTA4 (Glutathione S-Transferase Alpha 4)	6p12.2 52977948-52995380	17,433	Nkx2-5, Oct-B1, oct-B2, oct-B3, POU2F1, POU2F1a, POU2F2, POU2F2, POU2F2B, POU2F2C	Reactive electrophiles produced by oxidative metabolism have been linked to a number of degenerative diseases including Parkinson's disease, Alzheimer's disease, cataract formation, and atherosclerosis.
5	GSTA5 (Glutathione S-Transferase Alpha 5)	6p12.2 52831655-52846095	14,441	P-4 CUTL1, Evi-1, GATA-3, HNF-1, HNF-1A, HNF-4alpha1, YY1	Role in drug metabolism - cytochrome P450 and glutathione metabolism
6	GSTK1 (Glutathione S-Transferase Kappa 1)	7q34 143244093-143270854	26,762	AP-4, Cart-1, E47, Hand1, ISGF-3, p300, PPAR-gamma1, PPAR-gamma2, RFX1, STAT3	Localized in peroxisome and catalyzes the conjugation of glutathione to hydrophobic substates facilitating the removal of these compounds from cells
7	GSTM1 (Glutathione S-Transferase Mu 1)	1p13.3 109687796-109709039	21,244	AP-1, GR, GR-alpha, GR-beta	Detoxification of electrophilic compounds, including carcinogens, therapeutic drugs, environmental toxins and products of oxidative stress, by conjugation with glutathione
8	GSTM2 (Glutathione S-Transferase Mu 2)	1p13.3 109668022-109709551	41,530	AP-1, ATF-2, c-Jun, FOXO4, HNF-3beta, NF-kB, NF-kB1	Conjugation of reduced glutathione to a wide number of exogenous and endogenous hydrophobic electrophiles
9	GSTM3 (Glutathione S-Transferase Mu 3)	1p13.3 109733932-109741038	7,107	AREB6, LUN-1, NF-AT, NF-AT1, NF-AT2, NF-AT3, NF-AT4, NF-kB, NF-kB1, STAT5A	Govern uptake and detoxification of both endogenous compounds and xenobiotics at the testis and brain blood barriers

S. No.	Gene	Chromosomal/ Genomic location	Gene size (in bases)	Transcription Factors	Function
10	GSTM4 (Glutathione S-Transferase Mu 4)	1p13.3 109656076-109674836	18,761	FOXO4, HNF-3β, LyF-1, Meis-1, Meis-1a, Meis-1b, NF-kB NF-kB1, NF-kB2, YY1	Drug metabolism, cytochrome-P450 and Glutathione metabolism
11	GSTM5 (Glutathione S-Transferase Mu 5)	1p13.3 109711761-109775428	63,668	AML1a, AP-1, ARP-1, Chx10, HOXA3, LyF-1, Pax-2, Pax-2a, Pax-2b	Drug metabolism - cytochrome P450 and Glutathione metabolism
12	GSTO1 (Glutathione S-Transferase Omega 1)	10q25.1 104235356-104267464	32,109	Arnt, c-Ets-1, CREB delta, CREB, E47, GCNF, GCNF-1, GCNF-2, HEN1, Tal-1beta	Involved in the metabolism of xenobiotics and carcinogens like arsenate detoxification 1 (glutaredoxin) and Drug metabolism - cytochrome P450
13	GSTO2 (Glutathione S-Transferase Omega 2)	10q25.1 104268416-104304950	36,535	AREB6, c-Ets-1, CREB delta, CREB, FOXL1, HOXA9, HOXA9B, Meis-1, Meis-1a, p53	Drug metabolism - cytochrome P450 and Metabolism of water-soluble vitamins and cofactors
14	GSTP1 (Glutathione S-Transferase Pi 1)	11q13.2 67583595-67586656	3,062	AP-1, c-Jun, MyoD, Nkx2-5, p53, Sp1	Innate Immune System and Drug metabolism - cytochrome P450
15	GSTT1 (Glutathione S-Transferase Theta 1)	22q11.23 270308-278486	8,179	AML1a, STAT3	Catalyze the conjugation of reduced glutathione to a variety of electrophilic and hydrophobic compounds
16	GSTT2 (Glutathione S-Transferase Theta 2)	22q11.23 23980028-23983915	3,888	AP-1, ATF-2, c-Jun, GR, GR-alpha, GR-beta, p53, SREBP-1a, SREBP-1c	Catalyze the conjugation of reduced glutathione to a variety of electrophilic and hydrophobic compounds
17	GSTZ1 (Glutathione S-Transferase Zeta 1)	14q24.3 77320887-77331597	10,711	C/EBPalpha, HNF-3beta	Glutathione metabolism

Table 1. (Continued)

S. No.	Gene	Chromosomal/Genomic location	Gene size (in bases)	Transcription Factors	Function
18	MGST1 (Microsomal Glutathione S-Transferase 1)	12p12.3 16347142-16609259	262,118	FOXI1, HFH-3, POU2F1, POU2F1a	Production of leukotrienes and prostaglandin E, important mediators of inflammation
19	MGST2 (Microsomal Glutathione S-Transferase 2)	4q31.1 139665768-139740745	74,978	aMEF-2, AML1a, ATF-2, Brachyury Evi-1, FOXD1, GATA-2, HOXA5, MEF-2, MEF-2A	Catalyzes the conjugation of leukotriene A4 and reduced glutathione to produce leukotriene C4
20	MGST3 (Microsomal Glutathione S-Transferase 3)	1q24.1 165630873-165661796	30,924	AP-2, gamma Bach 2, C/EBPalpha, E47, GATA-1, GATA-6, HNF-1, HNF-1A, Pbx1a, RSRFC4	Glutathione-dependent peroxidase activity towards lipid hydroperoxides
21	PTGES (Prostaglandin E Synthase)	9q34.11 129738331-129777327	38,997	AML1a, STAT3	Eicosanoid Synthesis and Arachidonic acid metabolism.

The enzymatic detoxification of xenobiotics has been classified into three distinct phases which act in a tightly integrated manner. Phases I and II involve the conversion of a lipophilic, non-polar xenobiotic into a more water-soluble and therefore less toxic metabolite, which can then be eliminated more easily from the cell (phase III) (Sheehan et al., 2001). Phase I is catalysed mainly by the cytochrome P450 system. Phase II enzymes catalyse the conjugation of activated xenobiotics to an endogenous water-soluble substrate, such as reduced glutathione (GSH), UDP-glucuronic acid or glycine, dopamine, prostaglandins, chemotherapeutical substances, environmental carcinogens and products of lipid peroxidation and they are easily eliminated from the body (Nebert and Timothy, 2006). Those reactive metabolites which are not detoxified might react with DNA and lead to mutations (Figure 1). Low levels of phase II enzyme activity would therefore result in higher levels of active metabolites and consequently more DNA damage (Nebert and Timothy, 2006).

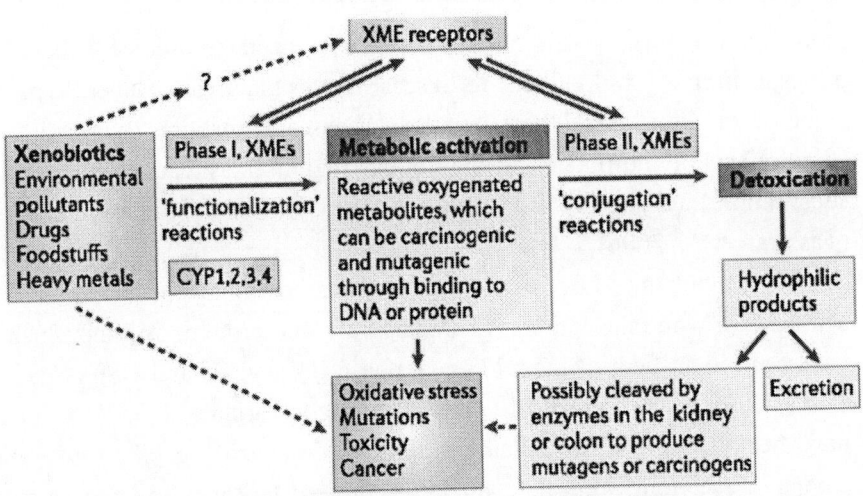

Figure 1. Classical scheme of phase I and phase II xenobiotic metabolizing enzymes (XMEs) and XME receptors. Certain environmental pollutants, drugs, foodstuffs and heavy metals interact with XME receptors or other reception pathways, leading to the upregulation or downregulation of XME gene expression (Adapted from Nebert and Dalton, 2006).

Quantitatively, conjugation to GSH, which is catalysed by GSTs, is the major phase II reaction in many species. GSTs catalyse nucleophilic aromatic substitutions and additions to α, β unsaturated ketones and epoxide ring-opening reactions, resulting in formation of oxidized glutathione (GSSG) (Hayes et al., 1999) by the formation of GSH conjugates (Salinas et al., 1999) and reduction of hydroperoxides (Figure 1).

GENETICS OF GLUTATHIONE *S*-TRANSFERASES

GSTs are encoded by 21 different genes which lie on different chromosome locations with definite promoter and specific transcription factors *viz.* *GSTA* (*A1-A5*) on 6p12.2, *GSTK1* on 7q34, *GSTM* (*M1-M5*) on 1p13.3, *GSTO* (*O1-O2*) on 10q25.1, *GSTP1* on 11q13, *GSTT* (*T1-T2*) on 22q11.2, *GSTZ1* on 14q24.3, *MGST1, T2, T3* on 12p12.3, 4q28 and 1q23 respectively (Table 1). Variants in genes coding for enzymes involved in protection against oxidative stress have been implicated in the predisposition of individuals to disease states such as cancer, type 2 diabetes mellitus and other complex diseases (Forsberg et al., 2001) (Figure 2). Polymorphisms in the coding regions may cause amino acid substitution thus altering the activity of xenobiotic metabolizing enzymes (Taskiran et al., 2006).

Several enzymes have been recognized as belonging to the Alpha and Mu classes, while the Phi class contains only one protein. Polymorphisms have been identified in several human phase II enzymes and GSTs (Hayes et al., 2000). In the case of Mu class, four allelic variants at *GST M1* locus have been identified in human population (Emahazion et al., 1999) and among these four, the null allele is present in 50% of the human population, which may predispose certain individuals to greater risk from toxic xenobiotics (Strange et al., 2001). A large number of Alpha-class GSTs, comprising at least six types of subunits have been identified in rat (Hsieh et al., 1997), with homologous gene loci in humans for A1-A4 (To-

Figueras et al., 2000). A number of polymorphisms have also been described in the Phi-class (Harris et al., 1998; Phuthong et al., 2018).

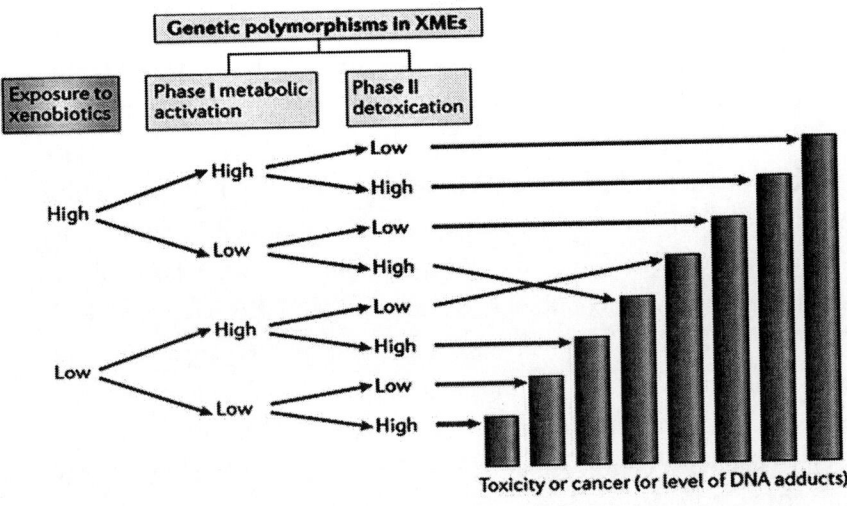

Figure 2. Effect of genetic polymorphisms in Xenobiotic Metabolizing Enzymes (XMEs) and the classical interaction of exposure with phase I and phase II XME metabolism, and risk of developing cancer (Adapted from Nebert and Dalton, 2006).

Theta-class GSTs have a catalytically essential serine rather than a tyrosine in the N terminus (Rossjohn et al., 1998; Mohana et al., 2017). It has been suggested that genes for enzymes similar to those in Theta class are widespread in nature, being found in bacteria, yeast, plants, insects and other sources (Tan et al., 2018). Two distinct homodimers (hGSTT1-1 and hGSTT2-2) have been identified in humans, showing only 50% sequence similarity (Bocedi et al., 2016). A null phenotype at the T1 locus in humans occurs in 10-38% of various ethnic groups which, as with the Mu class, may possibly underlie an increased risk of toxicity in response to certain xenobiotics (Hayes et al., 2000).

ROLE OF GSTS IN CERVICAL CANCER

Cervical cancer is the second common cancer in women between 15 and 44 years of age worldwide, with a prevalence estimated to be around 2.5 million, with some 569,847 new cases and 311,365 deaths worldwide (Ferlay et al., 2019). HPV types 16 and 18 are responsible for about 80% of all cervical cancer cases worldwide. Predisposing cofactors responsible for the development of cervical cancer are age, early marriage, number of abortions, young age at first delivery, parity, oral contraception, multiple sexual partners, smoking, low socio-economic status, menstrual hygiene and unhealthy living conditions (Sreedevi et al., 2015; Gupta et al., 2016). Life style and environment are two kinds of acquired susceptibility factors, while inherited factors are those genetic variants which are related to cervical cancer carcinogenesis. Both active as well as passive smoking habits have been confirmed to be risk factors for cervical cancer (Abbas et al., 2013). The presence of cotinine and nicotine in cervical mucus of women exposed to passive smoking can contribute to carcinogenesis through the same pathways as active smoking including genotoxic and immunomodulatory effects (Szarewski et al., 1996; Abbas et al., 2013). Although tobacco smoking is a risk factor and well established cause for cervical precancerous and cancerous states, it has not been explored to a great extent (Plummer et al., 2003). Active smokers are those who inhale smoke directly but passive smokers do not smoke themselves but inhale it from others smoking around them.

Cigarette smoking either active or passive has been linked to the secretion of tumor specific metabolites in cervical mucus. This mucus maintains cervical HPV infection longer and decreases potential of clearing an oncogenic infection (Giuliano et al., 2002). Chemical carcinogens are detoxified by phase II metabolizing enzymes e.g., glutathione S transferase (GST), N-acetyltransferase, epoxide hydroxylase and sulphotransferase (Lahdetie et al., 1986). GSTs belonging to mu, theta, and pi classes (*GSTM1*, *GSTT1* and *GSTP1*) play important roles in detoxification of metabolites of carcinogens in tobacco smoke (Parl, 2005). Polycyclic aromatic hydrocarbons and many other carcinogens require metabolic activation by phase I enzymes such as CYP1A1 which are detoxified by phase II enzymes such as GSTM1 and T1 (Oude Ophuis et al., 1998). Those reactive metabolites which are not detoxified might react with DNA and lead to mutations. Low levels of phase II enzyme activity would therefore result in higher levels of active metabolites and consequently more DNA damage.

GST POLYMORPHISMS AND CERVICAL CANCER

Genetic variations in genes encoding antioxidant enzymes (GSTs) have been observed amongst individuals which, lead to susceptibility and development of various cancers such as bladder, colon, gastrointestinal and lung cancers including cervical cancer (Katoh et al., 1996; Gronau et al., 2000). Environmental carcinogens such as 1, 3 butadiene found in air and tobacco smoke are detoxified by GSTT1 (Chen et al., 1999).

Null genotypes of *GSTM1* and *GSTT1* were found to be significantly associated with an increased risk of cervical cancer in Korean women and ~15-20% of Caucasians (Kim et al., 2000; Lee et al., 2004) including bladder, skin, lung, breast and ovary as well (Spurdle et al., 2001). An A/G polymorphism at the nucleotide level located within the substrate binding domain of GSTP1 leads to an amino acid variation of isoleucine to valine at codon 105 (Ile 105 Val) in the protein which decreases enzyme activity (Ali-Osman et al., 1997). Women having *GSTP1* 'AA' genotype have an increased risk of invasive cervical cancer which is higher among active smokers (Jee et al., 2002). In Indian population, women are rarely smokers however; exposure to smoke is passive which has been found to increase the risk of cervical cancer (Trimble et al., 2005; Abbas et al., 2013). The association of null genotypes of *GSTM1* (-/-) in passive smokers showed significant association (P = 0.004) with an increased risk of 4.19 folds of developing cervical cancer. In a previous study, *GSTP1* (AG) genotype showed a risk of 6.4-folds and significant association in case of passive smokers (Munoz et al., 2003) but study from our group showed an increased risk of 2.88 folds in passive smokers with *GSTP1* (GG) genotype while a marginal risk of 1.81 folds was observed in GSTT1 (-/-) individuals (Abbas et al., 2013).

A study from our group showed the association of *GST* polymorphisms with histological subtypes, squamous cell carcinoma and adenocarcinoma. The *GSTM1* (-/-) genotype was significantly (P = 0.02) associated with adenocarcinoma and the risk of developing cervical cancer was 4.01 folds. *GSTP1* genotypes did not show significant association with adenocarcinoma but individuals with 'AG' and 'GG' genotypes respectively were at 2.25 and 2.40 times higher risk of developing adenocarcinoma of the cervix (Abbas et al., 2013). Sobti et al., 2006, did not show any risk of developing cervical cancer with *GSTM1* (-/-) and T1 (-/-) genotypes taken together. Individuals with *GSTM1* (-/-) and *P1* (AG or GG) genotypes in double combination showed an increased risk of 2.98 folds while *GSTM1* (-/-), *T1* (+/+) and *P1* (AG or GG) revealed 3.04 folds increased risk and a significant association in triple combination (P = 0.02). *GSTM1* (-/-), *T1* (-/-) and *P1* (AA) showed non-significant association (P = 0.33) with cervical cancer but a risk of 2.36 folds, while *GSTM1* (-/-), *T1* (-/-) and *P1* (AG or GG) combination showed a 3.55-folds increased risk (Abbas et al., 2013).

GST POLYMORPHISMS AND CERVICAL CANCER TREATMENT

Cisplatin based concomitant chemoradiotherapy (CRT) is the standard treatment for cervical cancer which is known to improve survival and reduce the recurrence rate (Nogueira et al., 2012). Cisplatin is a platinum based drug used in the treatment of carcinoma of uterine cervix (Rotman et al., 1992). The platinum has the ability to cross-link the DNA molecules and generate intra-strand N-7 adducts which is major causes of cytotoxicity. During radiation, platinum compound forms free radicals with altered binding to DNA thus inhibiting DNA repair system (Coughlin et al., 1989). The cisplatin adducts become equated in tissue and interact with thiol containing molecules like glutathione and metallothioneins. Glutathione S-transferase (GST) detoxifies cisplatin by conjugation with glutathione and increases its excretion from the body (Rudin et al., 2003). Acquired resistance to cisplatin involves increased inactivation by glutathione and related enzymes (Siddik et al., 2003). This acquired chemoradioresistance is a serious clinical challenge which leads to recurrence of disease, progression and mortality (Lando et al., 2009).

Furthermore, the cytotoxic effects of radiation result principally from damage to DNA, either directly or indirectly by the formation of hydroxyl radicals and reactive oxygen species, which can be detoxified by GSTs (Tew et al., 1994). The cytotoxic effects of both radiation and chemotherapy are mediated primarily through generation of reactive oxygen species (ROS) and their byproducts causing cell damage. The tumor cell death by ROS is either by direct cytotoxic effect or intracellular apoptotic pathways. Therefore, various enzymes affecting ROS levels are likely to impact patient prognosis after treatment (Sun et al., 2013). Glutathione S-transferase (GST) is a phase II metabolic enzyme that plays an important role in cellular defense against numerous harmful chemicals produced both exogenously and endogenously (Tulsyan et al., 2013). It protects cells from oxidative damage by catalyzing conjugation of ROS (Yang et al., 2009) and also detoxifies various chemotherapeutic agents such as alkylating agents, platinum compounds and adriamycin (Ekhart et al., 2009). Glutathione (GSH) is one of the main detoxifying agents which can be affected by metabolic enzymes that use tripeptides as substrates. At the time of chemotherapy, sufficient activity of GST is crucial, since during the enzyme catalyzed conjugation reaction with GSH, the water solubility of drugs and other toxic materials increases and can be eliminated from the body (Economopoulos et al., 2010).

Polymorphisms in many genes contribute to significant treatment-related toxicities in patients by attenuating pathways like DNA repair, drug metabolism and cell cycle progression, impairing the survival of normal cells under stress during radiotherapy or chemotherapy. Genetic polymorphisms can affect protein expression and alter biological pathways that are integral in response to chemoradiation therapy in tumor cells

(Frazer et al., 2007). Many studies have shown that genetic variation in these enzymes, including *GSTP1* A313G (rs1695, Ile105Val) and whole gene deletions in *GSTM1* and *GSTT1* have impact on survival and toxicities in different cancers treated with platinum agents and radiation (Rednam et al., 2013). Due to the free radical scavenging activity and detoxification of platinum agents, role of GST in treatment response and survival in cervical cancer patients is plausible.

The direct involvement of GSTP1 in detoxification of cisplatin is by forming cisplatin-GSH adducts (Rolland et al., 2010). Many studies suggested that GSTs play an important role in development of tumor cell resistance to treatment. Reactive oxygen species (ROS) is generated during the treatment in both regimens *viz.* chemotherapy and radiotherapy. ROS is also generated as part of the cytotoxic activity of chemotherapeutic agents (Florea et al., 2001). Radiotherapy may kill cancer cells either directly through effect on target molecules or indirectly through ROS. These ROS can damage cells, proteins and DNA or other cellular molecules. Two possible mechanisms have been suggested for an association between GST genotypes and treatment outcome, one involving differences in carcinogen damage to DNA and the other, differences in detoxification by treatment agents or GST-mediated protection against oxidative damage during treatment (Sweeney et al., 2003).

In addition to GSTs, genetic polymorphisms in DNA repair genes are also known to be associated with cervical cancer (Abbas et al., 2019; Gupta et al., 2019) and treatment outcomes of CRT (Gurubhagavatula et al., 2004; Ryu et al., 2004). Many studies showed an association between polymorphisms of *GSTM1* and *T1* with treatment outcome in patients with breast cancer or childhood leukemia (Davies et al., 2001) but no association was found in colorectal cancer (Stoehlmacher et al., 2002). No individual effect of *GSTM1* or *T1* genotypes was found but in combination, null genotypes of both were associated with poor survival in ovarian cancer (Lallas et al., 2000). The study from our group reported that a reduced hazard of death and better overall survival was observed among CRT treated women with null genotype of *GSTM1* (M1-), particularly in combination with *GSTT1* null (M1/T1-) and *GSTP1* (M1-/AG+GG)

(Abbas et al., 2015). These results correspond with the hypothesis that treatment would be more successful among patients with less or no GST activity.

According to Stanulla et al., 2000, the risk of recurrence was reduced among children with *GSTM1* null, *GSTT1* null and *GSTP1* GG genotypes in acute lymphoblastic leukemia. This probably resulted in reduced or no enzymatic activity. We have also demonstrated that *GSTP1* AG+GG polymorphism individually did not show any association with survival but in combination with *GSTM1* null genotype (M1-/AG+GG) and both *GSTM1* and *GSTT1* null genotypes (M1-/T1-/AG+GG) showed reduced hazard ratio which means a better overall survival (Abbas et al., 2015).

Many studies link polymorphisms in *GSTs* to variation in cytotoxic effects of many chemotherapeutic drug and their association with survival and toxicity in many types of cancers *viz*. leukemia, lymphoma, glioma, breast, lung, ovarian, gastric, colorectal and germ cell tumors (Rolland et al., 2010). The gene deletion polymorphisms of *GSTM1* and *T1* have been described as null genotypes resulting in the absence of functional enzyme (Gurubhagavatula et al., 2004). A single nucleotide substitution (A>G) at position 313 leads to amino acid variation of isoleucine to valine at codon 105 (Ile105Val) of *GSTP1* which results in reduced enzymatic activity (Ryuet al., 2004). Genotypes resulting in lower GST activity may be advantageous for individuals undergoing chemoradiation treatment because a reduced detoxification may enhance the effectiveness of treatment (Cheng et al., 2009). Decreased enzyme activity due to deletion polymorphisms of *GSTM1* and *T1* increases treatment response as well as toxicity in patients receiving platinum-based drugs like cisplatin and oxaliplatin (Davies et al., 2001).

Our study has also demonstrated the impact of *GSTM1*, *GSTT1* and *GSTP1* gene variants on acute toxicity in cervical cancer patients undergoing CRT but did not find significant association (Abbas et al., 2018). We also reported that individuals having *GSTM1* null (-/-) genotype showed a higher risk of high grade gastrointestinal toxicity, but this was not statistically significant (Abbas et al., 2018). The enzyme product of GSTP1 is known to detoxify platinum compounds cisplatin and oxaliplatin, and *GSTP1* polymorphism is linked to differences in chemotherapy response and cancer susceptibility (McIlwain et al., 2006). In our study, patients having *GSTP1* AG or GG genotypes showed higher risk of high grade gastrointestinal toxicity, whilst the combination of genotypes *GSTM1* null/ *GSTP1* GG and GSTT1 null/*GSTP1* GG was linked to a significant higher risk of high grade gastrointestinal toxicity (Abbas et al., 2018).

GSTs AND OTHER CANCERS

There are evidences that GST polymorphisms underscore the relationship between GST isoforms conjugated GSH to electrophilic carcinogens leading to various other cancers such as those of the gastrointestinal tract, ovaries, prostate, and esophagus (Naidu et al., 2003; Akçay et al., 2005; Casson et al., 2006). Furthermore, the level of expression of GST could provide a useful diagnostic parameter in carcinoma of the breast (Forrester et al., 1990) and bladder (Engel et al., 2002). Evidence from molecular epidemiological studies showed that individual susceptibility to cancer is mediated by both genetic and environmental factors. Homozygous null genotypes of *GSTM1* and *GSTT1* genes are frequently found in lung and bladder cancers. Individuals with *GSTT1*- and *GSTM1*-null genotypes are more predisposed to acute myeloid leukemia (Crump et al., 2000) as well as oral leukoplakia risk as a result of carcinogenic intermediates derived from or generated during habitual chewing of betel quid/tobacco (Srivastava et al., 2011).

Bostwick et al. (2007) reported differential expression of *GSTA1*, *GSTM1* and *GSTP1* in benign prostate, prostatic intraepithelial neoplasia, and prostatic adenocarcinoma. The study observed that consistent reduction or loss of expression of all subclasses of GST could engender the progression of prostatic neoplasia from benign epithelium to high-grade prostatic intraepithelial neoplasia and carcinoma. Another study showed raised levels of *GSTA1*, *GSTM1* and *GSTP1* activities in confirmed oral epithelial dysplasias (OEDs) and squamous cell carcinomas in humans (Chen et al., 1997).

Another study also described the implication of total GST activity levels and GSTP1 protein expression in colorectal cancer, adenoma and normal mucosa (Naidu et al., 2003). It was shown that GST activity was raised significantly in both colorectal cancer and adenomas when compared with normal colonic tissue. It was indicative that raised levels of GST activity may serve as a useful diagnostic index for colonic neoplasia in humans.

Conversely, Szarka et al. (1995) noted low levels of GST activity in blood lymphocytes from high risk colorectal cancer patients in comparison to normal individuals ($P = 0.004$). However, no association was observed between the frequency of GSTM1 and risk of colorectal cancer. High-risk individuals were unable to express GSTM1 and had lower levels of GST activity when compared to control subjects ($P = 0.006$) (Szarka et al., 1995).

GSTs AND TYPE 2 DIABETES MELLITUS

Type 2 Diabetes Mellitus (T2DM) is a chronic disorder characterized by impaired metabolism of glucose and lipids due to defects in insulin secretion (beta cell dysfunction) or action (insulin resistance) (Vats et al., 2013). The characteristic properties of diabetes mellitus are chronic hyperglycemia, microvascular (e.g., retina, renal glomerulus and peripheral nerve) as well as macrovascular (e.g., atherosclerosis, coronary artery disease (CAD), stroke) pathologies (Banerjee and Vats, 2014) with more than 17.5 million deaths worldwide attributed to cardiovascular complications (IDF, 2015). This rise in the number of diabetic patients is associated with economic development, ageing population, increasing urbanization, dietary changes, reduced physical activity and changes in life style pattern (Moore et al., 2009).

T2DM and oxidative stress have both clinical and genetic correlation. The overall play of reactive metabolites (RMs) leads to the development of late onset insulin resistance (Banerjee and Saxena, 2014). RMs is generated inside the body of normal individuals in a scheduled manner and is in feedback control with the antioxidant system (Banerjee and Vats, 2014). The role of RMs and antioxidants in T2DM provides a lead for research in identifying antioxidant gene variants and risk genotypes in populations of different ethnicity (Vats et al., 2017). There are several risk variants of antioxidant enzyme genes for T2DM and associated complications (Banerjee and Vats, 2014). Recent advances of genome-wide studies have provided confirmation of association between common complex diseases and genetic variants. Antioxidant gene polymorphism studies are a comprehensive way of understanding the stress sensitive pathways.

GSTM1, T1 and *P1* have been reported to be involved in T2DM development and various diabetes related complications in different populations (Miller et al., 2003; Oniki et al., 2008; Bid et al., 2010; Amer et al., 2011; Cilenšek et al., 2012). Genetic variations in *GSTM1* can change an individual's susceptibility to carcinogens and toxins as well as affect the toxicity and efficacy of certain drugs. Pharmacogenetic studies provide insights on the relationship between individual genetic variants and variable therapeutic outcomes of various oral antidiabetic drugs (OADs). Clinical utility of pharmacogenetic study is to predict the therapeutic dose of various OADs on an individual basis. Pharmacogenetics therefore, is a step towards personalized medicine which will greatly improve the efficacy of diabetes treatment. These approaches focus on single nucleotide polymorphisms and their influence on individual drug response, efficacy and toxicity (Singh et al., 2016).

GST POLYMORPHISMS AND TYPE 2 DIABETES MELLITUS

In individual genotyping studies, there are reports of significant risk for *GSTM1* and *T1* null genotypes ($p<0.05$) in Turkish, Italian, North Indian, South Indian and Iranian populations (Bid et al., 2010; Yalin et al., 2007; Manfredi et al., 2009; Ramprasath et al., 2011; Moasser et al., 2012).

A non-significant association diabetes risk (NSDR) was reported in T2DM associated cardiovascular disease (CVD), diabetic retinopathy (DR) and diabetic nephropathy (DN) in mixed populations (Doney et al., 2005). Japanese and Egyptian subjects with T2DM showed association with high Triglyderide (TGA) and Low Density Lipoprotein (LDL), while Solvenian/Caucasian population with DR showed NSDR (Hori et al., 2007; Amer et al., 2011; Cilensek et al., 2012).

Individuals with *GSTM1* null genotype in Turkish population were at approximately 4 times higher risk (Yalinet al., 2007; Gonul et al., 2012), while Brazilian population reported an OR of 1.1 (Thameem et al., 2003).Similarly, Bid et al. (2010) and Mastana et al. (2013) reported null as the risk genotype with an Odds Ratio (OR) of 2.04 and 2.63 respectively. Iranian T2DM subjects showed OR of 1.74 (Moasser et al., 2012) and those with DR showed 1.43 (Moasser et al., 2014). Highly significant association (P<0.001) was reported in South Indian subjects with T2DM having OR of 2.92 (Ramprasath et al., 2011). Genotyping studies in North Indian population showed individuals with null genotype to be at 1.265 times higher risk (Vats et al., 2013).

The *GSTT1* gene is haplotype-specific and it is absent in 38% of the populations. Alternative splicing of this gene results in multiple transcript variants. *GSTT1* deletion polymorphism related reports showed significant, highly significant and non-significant association in various studies. Significant association (P<0.05) was found in Italian subjects with T2DM and associated coronary artery disease (CAD) with an OR indicating non-significant disease risk (NSDR) (Manfredi et al., 2009). Similarly, T2DM subjects showed significant association (P<0.05) with cardio-vascular disease (CVD) with OR of 2.7 in a mixed population study reported by Doney et al. (2005). However, *GSTT1* deletion polymorphism showed non-significant association with T2DM in Turkish population (Yalin et al., 2007). The T2DM associated diabetic retinopathy (DR) in Iranian population with OR of 1.41 (Moasser et al., 2014) and North Indian population with OR of 1.26 and 1.2 (Bid et al., 2010) were also reported. In a south Indian population study *GSTT1* null genotype has shown a highly significant association (p<0.001) with an OR = 3.11 (95%CI)

(Ramprasath et al., 2011). Similarly, a Brazilian population was at higher risk for the disease with OR = 3.2 (95%CI) (Pinheiro et al., 2013). In north Indian population there was an OR of 1.236 times although the individual genotypes, present or null were not found to be significant (Vats et al., 2013).

A *GSTP1* variant with a substitution in the active site, valine for isoleucine at codon 105 (Ile105Val) has a reduced ability to conjugate reactive electrophiles with glutathione and may therefore sensitize cells to free radical mediated damage. A cohort study carried out in Tayside region of Scotland showed no significant association with an OR of 1.04 (CI = 95%) (Doney et al., 2005). No association was observed in Turkish population (Yalin et al., 2007). However, a significant association (P<0.05) was reported in Caucasian population with significant disease risk (OR 1.25) (Cilensek et al., 2012). Likewise, in our study population a highly significant association (<0.0001) with an OR of 2.48 (CI = 95%) was observed (Vats et al., 2013).

The interaction of *GSTM1*(-/-), *GSTT1*(+/+) and *GSTP1*(A/A) together showed significant (P = 0.005) association with 2.43 times higher risk of T2DM (Bid et al., 2010). The interaction of *GSTM1*del/*GSTT1*del, NN showed significant association (p = 0.02) with 1.74 times risk while interaction of *GSTM1*(-/-), *GSTT1*(-/-) and *GSTP1*(A/G) i.e., NNV allele combination seemed to increase the risk up to 13.47 times (p = 0.001) (Banerjee et al., 2019).

Association of *SOD2*+47C/T polymorphism with T1DM was reported for the first time in Russian population and other populations as well (Nomiyama et al., 2003; Lee et al., 2006; de Jesús, et al., 2013). Although *SOD2*+47C/T alone showed only 0.64 times risk for T2DM and *GPx1* +599C/T showed no significant association (Vats et al., 2015), their interaction with *CAT*-21A/T i.e., TCC allele combination showed significant association (0.001) with 2.06 folds higher risk (Banerjee et al., 2019). No significant disease risk combination was observed during analysis of 04 SNPs in different genes. However, interaction of all gene variants taken together *viz.* *GSTM1*del *GSTT1*del *GSTP1*+313A/G (105Ile/Valrs1695) *CAT*-21A/T(rs7943316) *SOD2*+47C/T (rs4880) and

GPx1+599C/T (rs3811699) showed a very high degree up to 5083.34 times risk of developing T2DM in individual possessing 'PPITCT' allele combination (Banerjee et al., 2019).

CONCLUSION

Genetic studies will help to develop prognostic biomarkers for early prediction and risk assessment in patients and enable clinicians to develop personalized treatment regimes. This chapter represents advancement in biomedical science because it points to a number of genetic variants in GST encoding genes that together result in a very high risk of complex diseases. Therefore, it is important to study the polymorphic variants of important genes in different populations so that the risk genotypes can be identified which will be of clinical importance as indicators of disease susceptibility. Individuals at risk will be able to take prior preventive measures and delay the onset of disease. However, the complexity of clinical symptoms has made it difficult to develop therapies that provide a substantial improvement for extended periods of time in a wide range of patient groups.

ACKNOWLEDGMENTS

The authors are grateful to Indian Council of Medical Research (ICMR), Department of Science and Technology (DST), New Delhi, India and Centre of Excellence, Higher Education, Government of Uttar Pradesh, Lucknow, India for funding of genetic studies. Atar Singh Kushwah is thankful to ICMR, New Delhi for Senior Research Fellowship. Authors also acknowledge Dr. Pushpank Vats and Dr. Mohammad Abbas for their contribution in genotyping studies.

REFERENCES

Abbas M., K. Srivastava, M. Imran, and M. Banerjee. "Association of Glutathione S-transferase (GSTM1, GSTT1 and GSTP1) polymorphisms and passive smoking in cervical cancer cases from North India." *International Journal of Biomedical Research* 4 (2013): 655-62.

Abbas M., K. Srivastava, M. Imran, and M. Banerjee. "Genetic polymorphisms in DNA repair genes and their association with cervical cancer." *British Journal of Biomedical Science* 76, no. 3 (2019): 117-121.

Abbas Mohammad, V. S. Kushwaha, K. Srivastava, S. T. Raza, and M. Banerjee. "Impact of GSTM1, GSTT1 and GSTP1 genes polymorphisms on clinical toxicities and response to concomitant chemoradiotherapy in cervical cancer." *British Journal of Biomedical Science* 75, no. 4 (2018): 169-174.

Abbas M., V. S. Kushwaha, K. Srivastava, and M. Banerjee. "Glutathione S-transferase gene polymorphisms and treatment outcome in cervical cancer patients under concomitant chemoradiation." *PloS One* 10, no. 11 (2015).

Akçay T., Y. Dinçer, Z. Alademir, K. Aydınlı, M. Arvas, F. Demirkıran, and D. Kösebay. "Significance of the O6-methylguanine-DNA methyltransferase and glutathione S-transferase activity in the sera of patients with malignant and benign ovarian tumors." *European Journal of Obstetrics & Gynecology and Reproductive Biology* 119, no. 1 (2005): 108-113.

Ali-Osman F., O. Akande, G. Antoun, J. X. Mao, and J. Buolamwini. "Molecular cloning, characterization, and expression in *Escherichia coli* of full-length CDNAs of three human glutathione s-transferase pi gene variants evidence for differential catalytic activity of the encoded proteins." *Journal of Biological Chemistry* 272, no. 15 (1997): 10004-10012.

Amer M. A., M. H. Ghattas, D. M. Abo-Elmatty, and S. H. Abou-El-Ela. "Influence of glutathione S-transferase polymorphisms on type-2 diabetes mellitus risk." *Genetic Molecular Research* 10, no. 4 (2011): 3722-3730.

Arca P., C. Hardisson, and J. E. Suarez. "Purification of a glutathione S-transferase that mediates fosfomycin resistance in bacteria." *Antimicrobial Agents and Chemotherapy* 34, no. 5 (1990): 844-848.

Banerjee M., P. Vats, A. S. Kushwah, and N. Srivastava. "Interaction of antioxidant gene variants and susceptibility to type 2 diabetes mellitus." *British Journal of Biomedical Science* 76, no. 4 (2019): 166-171.

Banerjee M., and M. Saxena. "Genetic polymorphisms of cytokine genes in type 2 diabetes mellitus." *World Journal of Diabetes* 5, no. 4 (2014): 493.

Banerjee M., and P. Vats. "Reactive metabolites and antioxidant gene polymorphisms in type 2 diabetes mellitus." *Redox Biology* 2 (2014): 170-177.

Bid H. K., R. Konwar, M. Saxena, P. Chaudhari, C. G. Agrawal, and M. Banerjee. "Association of glutathione S-transferase (GSTM1, T1 and P1) gene polymorphisms with type 2 diabetes mellitus in north Indian population." *Journal of Postgraduate Medicine* 56, no. 3 (2010): 176.

Board G. P., T. Rohan Baker, G. Chelvanayagam, and S. L. Jermiin. "Zeta, a novel class of glutathione transferases in a range of species from plants to humans." *Biochemical Journal* 328, no. 3 (1997): 929-935.

Bocedi A., R. Fabrini, O. Lai, L. Alfieri, C. Roncoroni, A. Noce, J. Z. Pedersen, and G. Ricci. "Erythrocyte glutathione transferase: a general probe for chemical contaminations in mammals." *Cell Death Discovery* 2, no. 1 (2016): 1-5.

Bostwick D. G., I. Meiers, and J. H. Shanks. "Glutathione S-transferase: differential expression of α, μ, and π isoenzymes in benign prostate, prostatic intraepithelial neoplasia, and prostatic adenocarcinoma." *Human Pathology* 38, no. 9 (2007): 1394-1401.

Casson A. G, Z. Zheng, G. A. Porter, and D. L. Guernsey. "Genetic polymorphisms of microsomal epoxide hydrolase and glutathione S-transferases M1, T1 and P1, interactions with smoking, and risk for esophageal (Barrett) adenocarcinoma." *Cancer Detection and Prevention* 30, no. 5 (2006): 423-431.

Chen C., M. M. Madeleine, N. S. Weiss, and J. R. Daling. "Glutathione S-transferase M1 genotypes and the risk of squamous carcinoma of the cervix: a population-based case-control study." *American Journal of Epidemiology* 150, no. 6 (1999): 568-572.

Chen Y. K., and L-M. Lin. "Evaluation of glutathione S-transferase activity in human buccal epithelial dysplasias and squamous cell carcinomas." *International Journal of Oral and Maxillofacial Surgery* 26, no. 3 (1997): 205-209.

Cheng X-D, W-G. Lu, F. Ye, X-Y. Wan, and X. Xie. "The association of XRCC1 gene single nucleotide polymorphisms with response to neoadjuvant chemotherapy in locally advanced cervical carcinoma." *Journal of Experimental & Clinical Cancer Research* 28, no. 1 (2009): 91.

Cilenšek I., S. Mankoč, M. G. Petrovič, and D. Petrovič. "GSTT1 null genotype is a risk factor for diabetic retinopathy in Caucasians with type 2 diabetes, whereas GSTM1 null genotype might confer protection against retinopathy." *Disease Markers* 32, no. 2 (2012): 93-99.

Coughlin C. T., and R. C. Richmond. "Biologic and clinical developments of cisplatin combined with radiation: concepts, utility, projections for new trials, and the emergence of carboplatin." In *Seminars in Oncology* 16, no. 4 Suppl 6 (1989), pp. 31-43.

Crump C., C. Chen, F. R. Appelbaum, K. J. Kopecky, S. M. Schwartz, C. L. Willman, M. L. Slovak, and N. S. Weiss. "Glutathione S-transferase theta 1 gene deletion and risk of acute myeloid leukemia." *Cancer Epidemiology and Prevention Biomarkers* 9, no. 5 (2000): 457-460.

Davies S. M., L. L. Robison, J. D. Buckley, T. Tjoa, W. Woods, G. A. Radloff, J. A. Ross, and J. P. Perentesis. "Glutathione S-transferase polymorphisms and outcome of chemotherapy in childhood acute myeloid leukemia." *Journal of Clinical Oncology* 19, no. 5 (2001): 1279-1287.

de Jesús I. A-M, E. J. Parra, A. Valladares-Salgado, J. H. Gómez-Zamudio, J. Kumate-Rodriguez, J. Escobedo-de-la-Peña, and M. Cruz. "SOD2 gene Val16Ala polymorphism is associated with macroalbuminuria in Mexican Type 2 Diabetes patients: a comparative study and meta-analysis." *BMC Medical Genetics* 14, no. 1 (2013): 110.

Diabetes Atlas, 7th ed. International Diabetes Federation, Brussels, Belgium, http://www.diabetesatlas.org; 2015.

Dixon D. P., A. Lapthorn, and R. Edwards. "Plant glutathione transferases." *Genome Biology* 3, no. 3 (2002): reviews3004-1.

Doney S. F. A., S. Lee, G. P. Leese, A. D. Morris, and C. N. A. Palmer. "Increased Cardiovascular Morbidity and Mortality in Type 2 Diabetes Is Associated With the Glutathione S Transferase Theta–Null Genotype: A Go-DARTS Study." *Circulation* 111, no. 22 (2005): 2927-2934.

Economopoulos P. K., S. Choussein, N. F. Vlahos, and T. N. Sergentanis. "GSTM1 polymorphism, GSTT1 polymorphism, and cervical cancer risk: a meta-analysis." *International Journal of Gynecologic Cancer* 20, no. 9 (2010): 1576-1580.

Edwards R., D. P. Dixon, and V. Walbot. "Plant glutathione S-transferases: enzymes with multiple functions in sickness and in health." *Trends in Plant Science* 5, no. 5 (2000): 193-198.

Ekhart C., S. Rodenhuis, P. H. M. Smits, J. H. Beijnen, and A. D. R. Huitema. "An overview of the relations between polymorphisms in drug metabolising enzymes and drug transporters and survival after cancer drug treatment." *Cancer Treatment Reviews* 35, no. 1 (2009): 18-31.

Emahazion T., M. Jobs, W. M. Howell, M. Siegfried, P. I. Wyöni, J. A. Prince, and A. J. Brookes. "Identification of 167 polymorphisms in 88 genes from candidate neurodegeneration pathways." *Gene* 238, no. 2 (1999): 315-324.

Engel S. L., E. Taioli, R. Pfeiffer, M. Garcia-Closas, P. M. Marcus, Q. Lan, P. Boffetta. "Pooled analysis and meta-analysis of glutathione S-transferase M1 and bladder cancer: a HuGE review." *American Journal of Epidemiology* 156, no. 2 (2002): 95-109.

Ferlay J. M. Ervik, F. Lam, M. Colombet, L. Mery, M. Piñeros, A. Znaor, I. Soerjomataram, and F. Bray. "Global cancer observatory: cancer today." *Lyon, France: International Agency for Research on Cancer* (2018).

Florea A-M, and D. Büsselberg. "Cisplatin as an anti-tumor drug: cellular mechanisms of activity, drug resistance and induced side effects." *Cancers* 3, no. 1 (2011): 1351-1371.

Forrester M. L., J. D. Hayes, R. Millis, D. Barnes, A. L. Harris, J. J. Schlager, G. Powis, and C. R. Wolf. "Expression of glutathione S-transferases and cytochrome P450 in normal and tumor breast tissue." *Carcinogenesis* 11, no. 12 (1990): 2163-2170.

Forsberg L., U. deFaire, and R. Morgenstern. "Oxidative stress, human genetic variation, and disease." *Archives of Biochemistry and Biophysics* 389, no. 1 (2001): 84-93.

Frazer K. A., and Ballinger D. G. International HapMap Consortium. "A second generation human haplotype map of over 3.1 million SNPs." *Nature* 449, no. 7164 (2007): 851.

Fryer A., and R. Strange. "Glutathione S-transferase: genetics and role in toxicology." *Toxicology Letters* 112 (2000): 357-363.

Giuliano R. A., R. L. Sedjo, D. J. Roe, R. Harris, S. Baldwin, M. R. Papenfuss, M. Abrahamsen, and P. Inserra. "Clearance of oncogenic human papillomavirus (HPV) infection: effect of smoking (United States)." *Cancer Causes & Control* 13, no. 9 (2002): 839-846.

Gönül N., E. Kadioglu, N. A. Kocabaş, M. Özkaya, A. E. Karakaya, and B. Karahalil. "The role of GSTM1, GSTT1, GSTP1, and OGG1 polymorphisms in type 2 diabetes mellitus risk: a case–control study in a Turkish population." *Gene* 505, no. 1 (2012): 121-127.

Goodman T. M., K. McDuffie, B. Hernandez, C. C. Bertram, L. R. Wilkens, C. Guo, A. Seifried, J. Killeen, and L. LeMarchand. "CYP1A1, GSTM1, and GSTT1 polymorphisms and the risk of cervical squamous intraepithelial lesions in a multiethnic population." *Gynecologic Oncology* 81, no. 2 (2001): 263-269.

Gronau S., D. König-Greger, G. Rettinger, and H. Riechelmann. "GSTM1 gene polymorphism in patients with head and neck tumors." *Laryngo-Rhino-Otologie* 79, no. 6 (2000): 341-344.

Gupta K. Maneesh, A. S. Kushwah, R. Singh, and M. Banerjee. "Genotypic analysis of XRCC4 and susceptibility to cervical cancer." *British Journal of Biomedical Science* 77, no. 1 (2020): 7-12.

Gupta K. Maneesh, R. Singh, and M. Banerjee. "Cytokine gene polymorphisms and their association with cervical cancer: A North Indian study." *Egyptian Journal of Medical Human Genetics* 17, no. 2 (2016): 155-163.

Gurubhagavatula S., G. Liu, S. Park, W. Zhou, L. Su, J. C. Wain, T. J. Lynch, D. S. Neuberg, and D. C. Christiani. "XPD and XRCC1 Genetic Polymorphisms Are Prognostic Factors in Advanced Non-Small-Cell Lung Cancer Patients Treated With Platinum Chemotherapy." *Journal of Clinical Oncology* 22, no. 13 (2004): 2594-2601.

Harris M. J., M. Coggan, L. Langton, S. R. Wilson, and P. G. Board. "Polymorphism of the Pi class glutathione S-transferase in normal populations and cancer patients." *Pharmacogenetics* 8, no. 1 (1998): 27-31.

Hayes J. D., and D. J. Pulford. "The glutathione S-transferase supergene family: regulation of GST and the contribution of the Isoenzymes to cancer chemoprotection and drug resistance part I." *Critical Reviews in Biochemistry and Molecular Biology* 30, no. 6 (1995): 445-520.

Hayes J. D., and L. I. McLellan. "Glutathione and glutathione-dependent enzymes represent a co-ordinately regulated defence against oxidative stress." *Free Radical Research* 31, no. 4 (1999): 273-300.

Hayes J. D., and R. C. Strange. "Glutathione S-transferase polymorphisms and their biological consequences." *Pharmacology* 61, no. 3 (2000): 154-166.

Hori M., K. Oniki, K. Ueda, S. Goto, S. Mihara, T. Marubayashi, and K. Nakagawa. "Combined glutathione S-transferase T1 and M1 positive genotypes afford protection against type 2 diabetes in Japanese." *Pharmacogenomics* 8, no. 10 (2007): 1307.

Hsieh C. H., S. P. Tsai, H. I. Yeh, T. Chyisheu, and M. F. Tam. "Mass spectrometric analysis of rat ovary and testis cytosolic glutathione S-transferases (GSTs): identification of a novel class-alpha GST, rGSTA6*, in rat testis." *Biochemical Journal* 323, no. 2 (1997): 503-510.

Jee S. H., J. E. Lee, S. Kim, J. H. Kim, S. J. Um, S. J. Lee, S. E. Namkoong, and J. S. Park. "GSTP1 polymorphsim, cigarette smoking and cervical cancer risk in Korean women." *Yonsei Medical journal* 43, no. 6 (2002):712-716.

Kamisaka K., W. H. Habig, J. N. Ketley, I. M. Arias, and W. B. Jakoby. "Multiple forms of human glutathione S-transferase and their affinity for bilirubin." *European Journal of Biochemistry* 60, no. 1 (1975): 153-161.

Katoh T., N. Nagata, Y. Kuroda, H. Itoh, A. Kawahara, N. Kuroki, R. Ookuma, and D. A. Bell. "Glutathione S-transferase M1 (GSTM1) and T1 (GSTT1) genetic polymorphism and susceptibility to gastric and colorectal adenocarcinoma." *Carcinogenesis* 17, no. 9 (1996): 1855-1859.

Kim W. J., C. G. Lee, Y. G. Park, K. S. Kim, I. K. Kim, Y. W. Sohn, H. K. Min, J. M. Lee, and S. E. Namkoong. "Combined analysis of germline polymorphisms of p53, GSTM1, GSTT1, CYP1A1, and CYP2E1: relation to the incidence rate of cervical carcinoma." *Cancer: Interdisciplinary International Journal of the American Cancer Society* 88, no. 9 (2000): 2082-2091.

La R., D. Salome, and T. H. O. M. A. S. Leisinger. "Sequence analysis and expression of the bacterial dichloromethane dehalogenase structural gene, a member of the glutathione S-transferase supergene family." *Journal of Bacteriology* 172, no. 1 (1990): 164-171.

Lähdetie, J. "Micronucleated spermatids in the seminal fluid of smokers and nonsmokers." *Mutation Research/Genetic Toxicology* 172, no. 3 (1986): 255-263.

Lallas A. Thomas, S. K. McClain, M. S. Shahin, and R. E. Buller. "The glutathione S-transferase M1 genotype in ovarian cancer." *Cancer Epidemiology and Prevention Biomarkers* 9, no. 6 (2000): 587-590.

Lando M., M. Holden, L. C. Bergersen, D. H. Svendsrud, T.Stokke, K. Sundfør, I. K. Glad, G. B. Kristensen, and H. Lyng. "Gene dosage, expression, and ontology analysis identifies driver genes in the carcinogenesis and chemoradioresistance of cervical cancer." *PLoS Genetics* 5, no. 11 (2009).

Lee S-A, J. W. Kim, J. W. Roh, J. Y. Choi, K. M. Lee, K. Y. Yoo, Y. S. Song, and D. Kang. "Genetic polymorphisms of GSTM1, p21, p53 and HPV infection with cervical cancer in Korean women." *Gynecologic Oncology* 93, no. 1 (2004): 14-18.

Listowsky I., M. Abramovitz, H. Homma, and Y. Niitsu. "Intracellular binding and transport of hormones and xenobiotics by glutathionestransferases." *Drug Metabolism Reviews* 19, no. 3-4 (1988): 305-318.

Manfredi S., D. Calvi, M. D. Fiandra, N. Botto, A. Biagini, M. G. Andreassi. Glutathione S-transferase T1- and M1-null genotypes and coronary artery disease risk in patients with Type 2 diabetes mellitus. *Pharmacogenomics* 10, no. 1 (2009): 29-34.

Mastana S. S., A. Kaur, R. Hale, and M. R. Lindley. "Influence of glutathione S-transferase polymorphisms (GSTT1, GSTM1, GSTP1) on type-2 diabetes mellitus (T2D) risk in an endogamous population from north India." *Molecular Biology Reports* 40, no. 12 (2013): 7103-7110.

McIlwain C. C., D. M. Townsend, and K. D. Tew. "Glutathione S-transferase polymorphisms: cancer incidence and therapy." *Oncogene* 25, no. 11 (2006): 1639-1648.

McLellan L. I., and C. R. Wolf. "Glutathione and glutathione-dependent enzymes in cancer drug resistance." *Drug Resistance Updates* 2, no. 3 (1999): 153-164.

Miller A. E., J. S. Pankow, R. C. Millikan, M. S. Bray, C. M. Ballantyne, D. A. Bell, G. Heiss, and R. Li. "Glutathione-S-transferase genotypes, smoking, and their association with markers of inflammation, hemostasis, and endothelial function: the atherosclerosis risk in communities (ARIC) study." *Atherosclerosis* 171, no. 2 (2003): 265-272.

Moasser E., N. Azarpira, B. Shirazi, M. Saadat, and B. Geramizadeh. "Genetic polymorphisms of glutathione-s-transferase M1 and T1 genes with risk of diabetic retinopathy in Iranian population." *Iranian Journal of Basic Medical Sciences* 17, no. 5 (2014): 351.

Moasser E., S. R. Kazemi-Nezhad, M. Saadat, and N. Azarpira. "Study of the association between glutathione S-transferase (GSTM1, GSTT1, GSTP1) polymorphisms with type II diabetes mellitus in southern of Iran." *Molecular Biology Reports* 39, no. 12 (2012): 10187-10192.

Mohana Krishnamoorthy, and A. Achary. "Human cytosolic glutathione-S-transferases: quantitative analysis of expression, comparative analysis of structures and inhibition strategies of isozymes involved in drug resistance." *Drug Metabolism Reviews* 49, no. 3 (2017): 318-337.

Moore D. J., J. M. Gregory, Y. A. Kumah-Crystal, and J. H. Simmons. "Mitigating micro-and macro-vascular complications of diabetes beginning in adolescence." *Vascular Health and Risk Management* 5 (2009): 1015.

Morgenstern R., and J. W. DePierre. "Microsomal glutathione transferase: purification in unactivated form and further characterization of the activation process, substrate specificity and amino acid composition." *European Journal of Biochemistry* 134, no. 3 (1983): 591-597.

Muñoz N., F. X. Bosch, S. DeSanjosé, R. Herrero, X. Castellsagué, K. V. Shah, P. J. F. Snijders, and C. J. L. M. Meijer. "Epidemiologic classification of human papillomavirus types associated with cervical cancer." *New England Journal of Medicine* 348, no. 6 (2003): 518-527.

Naidu K. A., A. N., H. Pinkas, H. E. Kaiser, P. Brady, and D. Coppola. "Glutathione-S-transferase pi expression and activity is increased in colonic neoplasia." *In Vivo (Athens, Greece)* 17, no. 5 (2003): 479-482.

Nebert W. D., and T. P. Dalton. "The role of cytochrome P450 enzymes in endogenous signalling pathways and environmental carcinogenesis." *Nature Reviews Cancer* 6, no. 12 (2006): 947-960.

Nicholson W. D., A. Ali, J. P. Vaillancourt, J. R. Calaycay, R. A. Mumford, R. J. Zamboni, and A. W. Ford-Hutchinson. "Purification to homogeneity and the N-terminal sequence of human leukotriene C4 synthase: a homodimeric glutathione S-transferase composed of 18-kDa subunits." *Proceedings of the National Academy of Sciences* 90, no. 5 (1993): 2015-2019.

Nogueira A., R. Catarino, I. Faustino, C. Nogueira-Silva, T. Figueiredo, L. Lombo, I. Hilário-Silva, D. Pereira, and R. Medeiros. "Role of the RAD51 G172T polymorphism in the clinical outcome of cervical cancer patients under concomitant chemoradiotherapy." *Gene* 504, no. 2 (2012): 279-283.

Nomiyama T., Y. Tanaka, L. Piao, K. Nagasaka, K. Sakai, T. Ogihara, K. Nakajima, H. Watada, and R. Kawamori. "The polymorphism of manganese superoxide dismutase is associated with diabetic nephropathy in Japanese type 2 diabetic patients." *Journal of Human Genetics* 48, no. 3 (2003): 0138-0141.

Oniki K., Y. Umemoto, R. Nagata, M. Hori, S. Mihara, T. Marubayashi, and K. Nakagawa. "Glutathione S-transferase A1 polymorphism as a risk factor for smoking-related type 2 diabetes among Japanese." *Toxicology Letters* 178, no. 3 (2008): 143-145.

Oude Ophuis M. B., H. M. Roelofs, W. H. Peters, and J. J. Manni. "Glutathione S-transferase M1 and T1 and cytochrome P4501A1 polymorphisms in relation to the risk for benign and malignant head and neck lesions." *Cancer* 82, no. 5 (1998): 936-943.

Parl F. F. "Glutathione S-transferase genotypes and cancer risk." *Cancer Letters* 221, no. 2 (2005): 123-129.

Phuthong S., W. Settheetham-Ishida, S. Natphopsuk, and T. Ishida. "Genetic polymorphism of the glutathione S-transferase Pi 1 (GSTP1) and susceptibility to cervical cancer in human papilloma virus infected Northeastern Thai women." *Asian Pacific Journal of Cancer Prevention* 19, no. 2 (2018): 381.

Pinheiro S. D., C. R. R. Filho, C. A. Mundim, P. de M. Junior, C. J. Ulhoa, A. A. S. Reis, and P. C. Ghedini. "Evaluation of glutathione S-transferase GSTM1 and GSTT1 deletion polymorphisms on type-2 diabetes mellitus risk." *PLoS One* 8, no. 10 (2013).

Plummer M., R. Herrero, S. Franceschi, C. J. L. M. Meijer, P. Snijders, F. X. Bosch, S. deSanjosé, and N. Muñoz. "Smoking and cervical cancer: pooled analysis of the IARC multi-centric case–control study." *Cancer Causes & Control* 14, no. 9 (2003): 805-814.

Ranson H., La-aiedprapanthadara, and J. Hemingway. "Cloning and characterization of two glutathione S-transferases from a DDT-resistant strain of *Anopheles gambiae*." *Biochemical Journal*. 324, no. 1 (1997): 97-102.

Rednam S., M. E. Scheurer, A. Adesina, C. C. Lau, and M. F. Okcu. "Glutathione S-transferase P1 single nucleotide polymorphism predicts permanent ototoxicity in children with medulloblastoma." *Pediatric Blood & Cancer* 60, no. 4 (2013): 593-598.

Rolland D., M. Raharijaona, A. Barbarat, R. Houlgatte, and C. Thieblemont. "Inhibition of GST-pi nuclear transfer increases mantle cell lymphoma sensitivity to cisplatin, cytarabine, gemcitabine, bortezomib and doxorubicin." *Anticancer Research* 30, no. 10 (2010): 3951-3957.

Rossjohn J., W. J. McKinstry, A. J. Oakley, D. Verger, J.Flanagan, G. Chelvanayagam, K. L. Tan, P. G. Board, and M. W. Parker. "Human theta class glutathione transferase: the crystal structure reveals a sulfate-binding pocket within a buried active site." *Structure* 6, no. 3 (1998): 309-322.

Rotman M. Z. "Chemoirradiation: a new initiative in cancer treatment. 1991 RSNA annual oration in radiation oncology." *Radiology* 184, no. 2 (1992): 319-327.

Rudin M. C., Z. Yang, L. M. Schumaker, D. J. V. Weele, K. Newkirk, M. J. Egorin, E. G. Zuhowski, and K. J. Cullen. "Inhibition of glutathione synthesis reverses Bcl-2-mediated cisplatin resistance." *Cancer Research* 63, no. 2 (2003): 312-318.

Rushmore T. H., and C. B. Pickett. "Glutathione S-transferases, structure, regulation, and therapeutic implications." *Journal of Biological Chemistry* 268, no. 16 (1993): 11475-11478.

Ryu J-S, Y. C. Hong, H. S. Han, J. E. Lee, S. Kim, Y. M. Park, Y. C. Kim, and T. S. Hwang. "Association between polymorphisms of ERCC1 and XPD and survival in non-small-cell lung cancer patients treated with cisplatin combination chemotherapy." *Lung Cancer* 44, no. 3 (2004): 311-316.

Salinas E. A., and M. G. Wong. "Glutathione S-transferases-a review." *Current Medicinal Chemistry* 6, no. 4 (1999): 279-310.

Sheehan D., G. Meade, V. M. Foley, and C. A. Dowd. "Structure, function and evolution of glutathione transferases: implications for classification of non-mammalian members of an ancient enzyme superfamily." *Biochemical Journal* 360, no. 1 (2001): 1-16.

Siddik H. Z. "Cisplatin: mode of cytotoxic action and molecular basis of resistance." *Oncogene* 22, no. 47 (2003): 7265-7279.

Singh S., K. Usman, and M. Banerjee. "Pharmacogenetic studies update in type 2 diabetes mellitus." *World Journal of Diabetes* 7, no. 15 (2016): 302.

Sobti R. C., S. Kaur, P. Kaur, J. Singh, I. Gupta, V. Jain, and A. Nakahara. "Interaction of passive smoking with GST (GSTM1, GSTT1, and GSTP1) genotypes in the risk of cervical cancer in India." *Cancer Genetics and Cytogenetics* 166, no. 2 (2006): 117-123.

Spurdle B. A., P. M. Webb, D. M. Purdie, X. Chen, A. Green, and G. Chenevix-Trench. "Polymorphisms at the glutathione S-transferase GSTM1, GSTT1 and GSTP1 loci: risk of ovarian cancer by histological subtype." *Carcinogenesis* 22, no. 1 (2001): 67-72.

Sreedevi A., R. Javed, and A. Dinesh. "Epidemiology of cervical cancer with special focus on India." *International Journal of Women's Health* 7 (2015): 405.

Srivastava R., R. Sharma, S. Mishra, and R. B Singh. "Biochemical and molecular biological studies on oral cancer: An overview." *The Open Nutraceuticals Journal.* 4, no. 1 (2011).

Stanulla M., M. Schrappe, A. M. Brechlin, M. Zimmermann, and K. Welte. "Polymorphisms within glutathione S-transferase genes (GSTM1, GSTT1, GSTP1) and risk of relapse in childhood B-cell precursor acute lymphoblastic leukemia: a case-control study." *Blood, the Journal of the American Society of Hematology* 95, no. 4 (2000): 1222-1228.

Stoehlmacher J., D. J. Park, W. Zhang, S. Groshen, D. D. Tsao-Wei, M. C. Yu, and H. J. Lenz. "Association between glutathione S-transferase P1, T1, and M1 genetic polymorphism and survival of patients with metastatic colorectal cancer." *Journal of the National Cancer Institute* 94, no. 12 (2002): 936-942.

Sun X., M. Ai, Y. Wang, S. Shen, Y. Gu, Y. Jin, Z. Zhou, Y. Long, and Q. Yu. "Selective induction of tumor cell apoptosis by a novel P450-mediated reactive oxygen species (ROS) inducer methyl 3-(4-nitrophenyl) propiolate." *Journal of Biological Chemistry* 288, no. 13 (2013): 8826-8837.

Sweeney C., V. Nazar-Stewart, P. L. Stapleton, D. L. Eaton, and T. L. Vaughan. "Glutathione S-transferase M1, T1, and P1 polymorphisms and survival among lung cancer patients." *Cancer Epidemiology and Prevention Biomarkers* 12, no. 6 (2003): 527-533.

Szarewski A., P. Sasieni, R. Edwards, J. Cuzick, M. J. Jarvis, Stuart J. Steele, and J. Guillebaud. "Effect of smoking cessation on cervical lesion size." *The Lancet* 347, no. 9006 (1996): 941-943.

Szarka E. C., G. R. Pfeiffer, S. T. Hum, L. C. Everley, A. M. Balshem, D. F. Moore, S. Litwin. "Glutathione S-transferase activity and glutathione S-transferase μ expression in subjects with risk for colorectal cancer." *Cancer Research* 55, no. 13 (1995): 2789-2793.

Tan H. M., and W. Y. Low. "Rapid birth-death evolution and positive selection in detoxification-type glutathione S-transferases in mammals." *PloS One* 13, no. 12 (2018).

Tew Kenneth D. "Glutathione-associated enzymes in anticancer drug resistance." *Cancer Research* 54, no. 16 (1994): 4313-4320.

Thameem F., X. Yang, P. A. Permana, Johanna K. Wolford, Clifton Bogardus, and Michal Prochazka. "Evaluation of the microsomal glutathione S-transferase 3 (MGST3) locus on 1q23 as a Type 2 diabetes susceptibility gene in Pima Indians." *Human Genetics* 113, no. 4 (2003): 353-358.

Tharmarajan R., P. S. Murugan, A. D. Prabakaran, P. Gomathi, A. Rathinavel, and G. S. Selvam. "Potential risk modifications of GSTT1, GSTM1 and GSTP1 (glutathione-S-transferases) variants and their association to CAD in patients with type-2 diabetes." *Biochemical and Biophysical Research Communications* 407, no. 1 (2011): 49-53.

To-Figueras J., M. Gene, J. Gomez-Catalan, E. Pique, N. Borrego, G. Marfany, R. Gonzalez Duarte, and J. Corbella. "Polymorphism of glutathione S-transferase M3: interaction with glutathione S-transferase M1 and lung cancer susceptibility." *Biomarkers* 5, no. 1 (2000): 73-80.

Trimble L. C., J. M. Genkinger, A. E. Burke, S. C. Hoffman, K. J. Helzlsouer, M. Diener-West, G. W. Comstock, and A. J. Alberg. "Active and passive cigarette smoking and the risk of cervical neoplasia." *Obstetrics and Gynecology* 105, no. 1 (2005): 174.

Tulsyan S., P. Chaturvedi, G. Agarwal, P. Lal, S. Agrawal, R. Devi Mittal, and B. Mittal. "Pharmacogenetic influence of GST polymorphisms on anthracycline-based chemotherapy responses and toxicity in breast cancer patients: a multi-analytical approach." *Molecular Diagnosis & Therapy* 17, no. 6 (2013): 371-379.

Vats P., N. Sagar, T. P. Singh, and M. Banerjee. "Association of superoxide dismutases (SOD1 and SOD2) and glutathione peroxidase 1 (GPx1) gene polymorphisms with type 2 diabetes mellitus." *Free Radical Research* 49, no. 1 (2015): 17-24.

Vats P., A. S. Kushwah, and M. Banerjee. "Association of antioxidant gene variants with type 2 diabetes mellitus in different ethnic groups." *European Journal of Biomedical and Pharmaceutical Sciences* 4 (2017): 290-298.

Vats P., H. Chandra, and M. Banerjee. "Glutathione S-transferase and Catalase gene polymorphisms with Type 2 diabetes mellitus." *Disease and Molecular Medicine* 1, no. 3 (2013): 46-53.

Yalin S., R. Hatungil, L. Tamer, N. A. Ates, N. Dogruer, H. Yildirim, S. Karakas, and U. Atik. "Glutathione S-transferase gene polymorphisms in Turkish patients with diabetes mellitus." *Cell Biochemistry and Function: Cellular Biochemistry and Its Modulation by Active Agents or Disease* 25, no. 5 (2007): 509-513.

Yang Y., K. K. Parsons, L. Chi, S. M. Malakauskas, and T. H. Le. "Glutathione S-transferase-μ1 regulates vascular smooth muscle cell proliferation, migration, and oxidative stress." *Hypertension* 54, no. 6 (2009): 1360-1368.

In: Glutathione S-Transferases
Editor: Igor Azevedo Silva

ISBN: 978-1-53618-188-3
© 2020 Nova Science Publishers, Inc.

Chapter 3

ANTIOXIDANT ACTIVITY OF GLUTATHIONE S-TRANSFERASES AND THEIR ASSOCIATION WITH BLADDER CANCER

Md. Bayejid Hosen and Yearul Kabir, PhD
Department of Biochemistry and Molecular Biology,
University of Dhaka, Dhaka, Bangladesh

ABSTRACT

Glutathione S-transferases (GSTs) are phase II metabolic enzymes which participate in the cellular detoxification processes against xenobiotics and noxious compounds as well as against oxidative stress. According to cellular compartment, the GSTs are divided into three GSTs superfamilies. The cytoplasmic (cGSTs), mitochondrial (κGSTs), and microsomal (also known as MAPEG) GSTs. Among these families, cGSTs are the most complex and most linked to human diseases, and according to similarities in amino-acid sequences, different structure of genes, and immunological cross-reactivity, cGSTs are divided into seven subtypes (α, π, μ, θ, ω, σ, and δ). GSTs act as antioxidant by conjugating glutathione with toxic molecules and detoxify a wide range of endogenous and environmental reactive oxygen species (ROS) such as superoxide, nitric oxide, hydroxyl radicals, ethylene oxide, polycyclic aromatic hydrocarbon epoxide present in tobacco smoke and other reactive species (RS) such as hydrogen peroxide, peroxynitrite, hypochlorous acid. There are different kinds of ROS and RS which act as carcinogens and cause damage to DNA, RNA, proteins and lipids and involved in the mutagenesis process, thereby resulting in human disease like bladder cancer. Bladder cancer is one of the most common cancers of the urinary tract which is said to be affected by occupational chemicals, for instance, tobacco, 2-naphthylamine, benzidine, and 4-aminobiphenyl etc. When cellular detoxifying enzymes are less active or malfunctioned, these chemicals escape the detoxifying reactions, and hence deposited in the bladder, thereby causing damage to bladder wall which initiate bladder tumorigenesis. It is evident from numerous studies that GSTs scavenge bladder tumor formation by detoxifying these chemicals and facilitating their excretion through urine without any damage to bladder epithelial cells. It is evidenced that single nucleotide polymorphisms of different GSTs superfamilies are associated with the occurrence of bladder cancer. Though the expression of GSTs is essential for eradication of toxic chemicals and tumor cells, overexpression of GSTs have also been associated with survival of tumor cells. Overexpression of GSTs cause neutralization of cellular ROS which are indispensable for apoptosis of tumor cells and detoxify anticancer drugs that are used for cancer treatment. It has also been shown that, some GSTs enzymes are exceptionally overexpressed in tumor cells and study of GSTs has introduced different anticancer drugs such as GST inhibitors, glutathione analogues, pro-drugs etc. and also new scaffolds and analogues are reported every year. This chapter will discuss all of these issues in depth focusing on the antioxidant activities of GSTs and their association with bladder cancer.

INTRODUCTION

All aerobic organisms, including bacteria, plants, insects and mammals, are subject to oxidative stress resulting from over production of reactive oxygen species (ROS), such as free radicals (e.g., superoxide, nitric oxide (NO) and hydroxyl radicals (OHs) and other reactive species (RS) (e.g., hydrogen peroxide, peroxynitrite and hypochlorous acid) (Lukasik et al., 2007). ROS are produced endogenously as a result of aerobic metabolisms, such as oxygen-consuming respiration in the mitochondria and oxygen-producing photosynthesis in the chloroplast. The

ROS produced may react with biomolecules, such as DNA, RNA, proteins and lipids, resulting in cell damage and death (Apel et al., 2004). Moreover, aerobic organisms also face toxic challenges from various exogenous toxic and/or pro-oxidant xenobiotics present in their diet or environment. Xenobiotics may exert toxic effects by directly disrupting the normal functions of the organisms and/or by generating ROS upon entry to the body or oxidation (Ahmed, 1992; Reed, 1995).

To cope with the inevitable oxidative stress from aerobic metabolisms, toxin-rich diets or the environment, aerobic organisms have evolved an integrated detoxification system capable of scavenging ROS and degrading toxic xenobiotics, as well as complex regulatory machinery capable of inducing certain components of the detoxification system when encountering a particular ROS or xenobiotic (Apel et al., 2004; Li et al., 2007). This detoxification system heavily depends upon reduced glutathione (γ-glutamyl-cysteinyl-glycine; GSH), an endogenous antioxidant molecule, and glutathione-S-transferase (GST), an antioxidant/detoxification enzyme. GSH and GST are ubiquitous in all living organisms and often exert synergistic actions in detoxifying both ROS and xenobiotics (Li et al., 2011).

Bladder cancer, also known as urothelial cancer of the bladder, is the most common malignancy affecting the urinary system (Chen et al., 2008; John et al., 2017; Zhang et al., 2017). Treatment of bladder cancer has not been advanced in the past 30 years (John et al., 2017). The disease has a multifactorial etiology that includes environmental factors such as cigarette smoking, arsenic exposure and occupational exposure as well as genetic factors (Chen et al., 2009; Oliveira et al., 2012; Reszka et al., 2012). Genetic factors are the one of the most important factors associated with the onset of bladder cancer (Benhamou et al., 2016). Smoking is a major risk factor for the development of this cancer, but the functional consequences of the carcinogens in tobacco smoke in terms of bladder cancer–associated metabolic changes remain poorly defined. Current evidence indicates that some gene polymorphisms are associated with bladder cancer morbidity (Selinski et al., 2017a; Selinski et al., 2017b; Singh et al., 2014; Verma et al., 2017; Wieczorek et al., 2014). It is also evident from numerous studies that GSTs scavenge bladder tumor formation by detoxifying carcinogenic chemicals and facilitating their excretion through urine without any damage to bladder epithelial cells (Albarakati et al., 2019; Huang et al., 2015). By the advances of cutting-edge technologies, it has already been established that point mutations and polymorphisms of different GSTs superfamilies are strongly associated with the occurrence of bladder cancer (Albarakati et al., 2019; Wang et al., 2013; Wu et al., 2013). Studies during the past decades have shed some light on the effect of GSTs in bladder cancer risk and our knowledge about the impact of various GSTs in the susceptibility, drug resistance and apoptosis of bladder tumors facilitate the improvement of the treatment of patients with bladder cancer. This chapter presents the current understanding of GSH and GST enzymes, with emphasis on their roles in detoxification, their reaction mechanisms as well as molecular mechanisms of GSTs for implication of bladder cancer.

ANTIOXIDANT ACTIVITIES OF GLUTATHIONE AND GLUTATHIONE-S-TRANSFERASES

Glutathione (GSH)

GSH is the most abundant intracellular nonprotein thiol (RSH) molecule. Its high electron-donating sulfhydryl (-SH) group on the cysteinyl portion combined with high intracellular concentration (millimolar levels) generate great reducing power. This characteristic allows GSH to participate in a wide variety of metabolic protection and/or detoxification processes, including oxidation-reduction (redox) reactions and nucleophilic displacement or addition-type conjugation reactions (Franco et al., 2007; Wu et al., 2004). When participating in redox reactions, GSH acts as an antioxidant, and scavenges a wide range of RS, including ROS and reactive nitrogen species (RNS), which are generated endogenously or from exogenous toxic and/or pro-oxidant compounds.

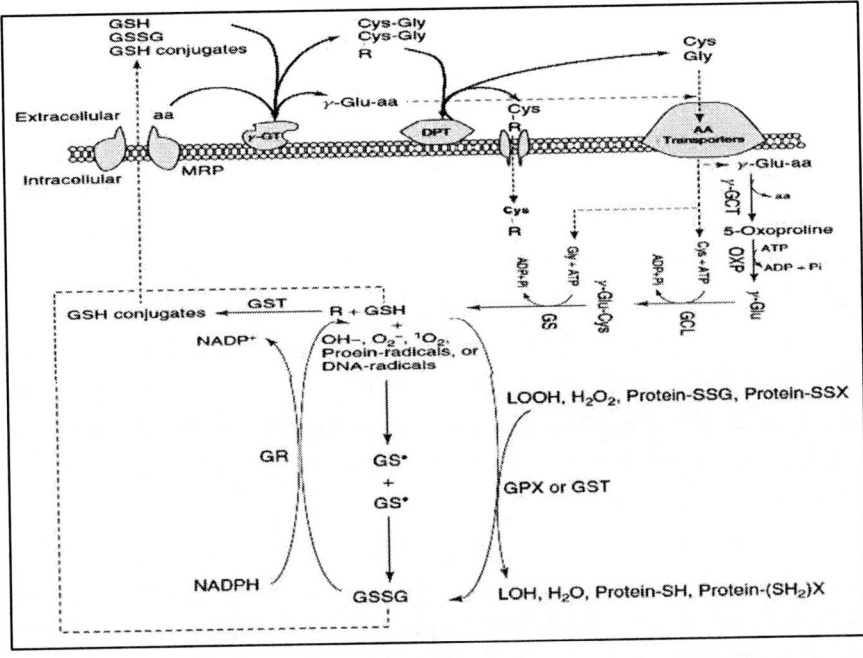

Figure 1. GSH biosynthesis, its utilization and recycling from GSSG and GSH conjugates. GSH is synthesized within the cytoplasm in a two-step enzymatic process catalysed by glutamyl cysteine ligase (GCL) and GSH synthetase (GS). GSH can directly or catalytically (by GSH peroxidase (GPX) or glutathione-S-transferase (GST)) reduce various reactive species and oxidants through formation of glutathione disulfide (GSSG). GSH can also conjugate with endogenous and exogenous electrophiles (R), which is largely catalyzed by GST. While GSSG can be reduced enzymatically by the NADPH-dependent glutathione reductase (GR) to regenerate GSH within the cells, GSH conjugates must be exported into the extracellular space by ATP-dependent multidrug-resistance protein (MRP) transporters to avoid deleterious effects. MRP may export some GSSG to the extracellular space as well, when the cellular GSSG concentration is too high. The extracellular GSH conjugates, GSH and GSSG are metabolized by γ-glutamyl transpeptidase (γ-GT) and dipeptidase (DPT) to γ-glutamyl-amino acid (γ-Glu-aa), cysteine (Cys), glycine (Gly) and cysteine conjugates (Cys R, breakdown products of GSH conjugates). The first three products are transported back into the cells by specific amino-acid transporters to serve as substrates for GSH biosynthesis. Metabolic pathways are depicted by solid lines, whereas pathways showing transport are dashed (Adapted from Li, 2011).

As an antioxidant, GSH can directly react with NO to form the relatively stable S-nitrosoglutathione (GSNO) (GSH + NO + O_2 → GSNO + O_2-), which can be inactivated by its conjugation to proteins through S-nitrosylation reactions (GSNO + Protein-SH → Protein-SNO + GSH). GSH can also effectively scavenge a number of ROS, including the superoxide anion (•O_2^-), •OH^-, singlet oxygen ($^1O_2$) and protein and DNA radicals directly, by donating electrons and becoming oxidized to the thiyl radical (GS•) (Figure 1) (Franco et al., 2007). Moreover, GSH can function as an essential cofactor for GSH peroxidases (GPXs), peroxiredoxins (PXRs), phospholipid hydroperoxide GSH peroxidases (PHGPXs) and GST, to catalytically detoxify cells from hydroperoxides (e.g., H_2O_2), peroxynitrite ($OONO^-$) and lipid peroxides (LOOH) (Figure 1) (Mates, 2000; Valko et al., 2007).

In these direct and indirect (enzyme-catalyzing) redox reactions, RS are reduced or inactivated mainly through the generation of disulfide bonds between two GS molecules to form glutathione disulfide (GSSG) (Figure 1). This disulfide linkage is reversible and one GSSG molecule can be rapidly reduced to two GSH molecules by the NADPH-dependent glutathione reductase (GR) (Figure 1). Changes in the intracellular thiol–disulfide (GSH/GSSG) balance within the cell can be used as an indicator of the redox status of the cells. Under normal physiological conditions, the resulting GSSG can be rapidly reduced to GSH by GR (Figure 1), keeping GSSG at ~1% of total GSH concentration (Forman et al., 2002; Wu et al., 2004).

When participated in nucleophilic displacement or addition-type conjugation reactions, GSH plays an essential role in the detoxification and elimination of a great variety of exogenous toxic xenobiotics (e.g., arene oxides, unsaturated carbonyls, organic halides), drugs and endogenous electrophiles (e.g., oestrogen, melanins, prostaglandins and leukotrienes) (Franco et al., 2007). In these conjugation reactions, the sulfhydryl group of GSH acts as a nucleophile and either displaces another atom (e.g., Cl) or group (e.g., nitro) or attacks an electrophilic site in xenobiotics, endogenous compounds or their reactive metabolism intermediates, to form GSH conjugates (Figure 1) (Blair, 2006; Franco et al., 2007). GSH conjugation reactions may be spontaneous, but primarily are catalyzed by GST, a family of Phase II detoxification enzymes (Wu et al., 2004). To some substrates, GSH conjugation can retain activity of the parent compounds or even activate the parent compounds (Blair, 2006). For example, directly toxic GSH conjugates may be formed from vicinal dihaloalkanes via formation of electrophilic sulphur mustards. Precursor organic thiocyanates and nitrosoguanidines (N-methyl-N'-nitro-N-nitroguanidine) may release toxic agents on GSH conjugation (Monks et al., 1990). But, in general, GSH conjugation is considered as an important detoxification mechanism because most of the GSH conjugates produced are more water soluble and less toxic than the parent electrophiles.

Glutathione-S-Transferases (GSTs)

The functions of GSTs are extremely versatile; this is manifested by the types of reactions they catalyze, the diversity of their substrates, and their noncatalytic activities. As a superfamily of Phase II detoxification enzymes, GSTs function primarily as GSH transferases to catalyze the conjugation of electrophilic endobiotics and xenobiotics (or their metabolites) to GSH (Figure 1) (Blanchette et al., 2007; Hayes et al., 2005; Li et al., 2007; Ranson and Hemingway, 2005; Sharma et al., 2004). Some GSTs also function as selenium (Se)-independent GSH peroxidase (Figure 1), GSH-dependent isomerase and GSH-dependent DDT dehydrochlorinase (DDTase). In addition, GSTs also have noncatalytic activities, functioning as binding and carrier proteins i.e., ligadins of various toxic compounds, modulators of the mitogen-activated protein (MAP) kinase signaling pathway via protein–protein interactions (Blanchette et al., 2007; Hayes et al., 2005; Sharma et al., 2004). As discussed below, most of these activities are essential for cell survival and detoxification against oxidants, RS and toxic electrophiles.

GST plays a pivotal role in cellular protection and detoxification by functioning as GSH transferases. As described in Figure 1, GSH confers protection by participating in two types of metabolic reactions: redox reactions for inactivating various oxidants/RS and scavenging the free radicals, and conjugation reactions for detoxifying toxic electrophilic compounds. Both types of reactions, in most cases, require catalysis of antioxidant/detoxificative enzymes. GSTs are the only enzymes that catalyze GSH conjugation reactions with endogenous and exogenous electrophiles. And GSH conjugation is a key stage in the conversion of lipophilic compounds to water-soluble metabolites that are more readily eliminated from the body (Ranson and Hemingway, 2005). Catalysis of GSH conjugations with various electrophiles is the basic function of GSTs. Almost all GSTs can function as GSH transferases although different GSTs may differ in the substrate specificities. Endogenous harmful electrophiles detoxified by GST-mediated GSH conjugations include epoxides, leukotrienes, the arachidonic acid oxidation products oxyeicosanoids (15-deoxy-Δ12,14-prostaglandin J2), the quinone-containing metabolites (aminochrome, dopachrome, noradrenochrome and adrenochrome) from catecholamine oxidation, the lipid peroxidation products, α, β-unsaturated aldehydes (2-alkenals, acrolein, crotonaldehyde, 4-hydroxy-2-alkenals, 4-hydroxy-2(E)-nonenal, 4-oxo-2(E)-nonenal), the cholesterol peroxidation products, cholesterol-5,6-oxide, epoxyeicosatrienoic acid and 9,10-epoxystearic acid, the oestradiol catechol metabolites, 2-hydroxy-oestradiol and 4-hydroxy-oestradiol and oxidized DNA bases (Blair, 2006; Hayes et al., 2005). The GSH conjugates formed are further metabolized and finally excreted as mercapturic acid derivatives in mammals (Figure 1).

A wide range of exogenous xenobiotics and/or their Phase I metabolites are detoxified by GST-mediated GSH conjugations. These include phytochemicals, therapeutic drugs, industrial intermediates, insecticides, herbicides, environmental pollutants and carcinogens (Hayes et al., 2005; Li et al., 2007; Ranson and Hemingway, 2005). Dietary phytochemicals that are metabolized by GST-mediated GSH conjugation reactions in animals include flavonoids, organothiocyanates (allyl thiocyanate, benzyl thiocyanate and 2-phenylethyl isothiocyanate) and α, β-unsaturated carbonyls (trans-2-octenal, trans-2-nonenal, 2, 4-hexadienal, trans, trans-2, 4-heptadienal, trans,trans-2, 4-nonadienal and trans,trans-2,

4-decadienal) (Li et al., 2007). Adriamycin, 1,3-bis(2-chloroethyl)-1-nitrosourea (BCNU), busulfan, carmustine, chlorambucil, cisplatin, crotonyloxymethyl-2-cyclohexenone (COMC-6), cyclophosphamide, ethacrynic acid, melphalan, mitozantrone and thiotepa are among the cancer chemotherapeutic drugs detoxified by GSTs (Hayes et al., 2005). GST mediated GSH conjugations are implicated in the detoxification and resistance of some of the organophosphate insecticides and several herbicides including atrazine, fluorodifen, pentachloronitrobenzene (PCNB), propachlor, chlorimuron ethyl (Li et al., 2007; O¨ztetik, 2008). A large number of epoxides, such as the antibiotic fosfomycin and those formed from environmental carcinogens aflatoxin B1, 1-nitropyrene, 4-nitroquinoline, polycyclic aromatic hydrocarbons (PAHs) and styrene, by the actions of cytochrome P450 monooxygenases in the liver, lung, gastrointestinal tract and other organs, are detoxified by GSTs (Hayes et al., 2005). GSTs have been also documented to detoxify the ultimate carcinogenic bay- and fjord-region diol epoxides produced from chrysene, methylchrysene, benzo[c]chrysene, benzo[g]chrysene, benzo[c]phenanthrene, benzo[a]pyrene, dibenz[a,h]anthracene and dibenzo[a,l]pyrene, as well as the carcinogenic heterocyclic amines found in cooked protein-rich food (Hayes et al., 2005).

In addition to catalyzing the conjugation of electrophilic substrates to GSH, many GST enzymes, such as the mitochondrial kappa (K) class GSTs and the cytosolic alpha (A) class GSTs also have a high Se-independent GSH peroxidase activity that allows them to catalyze some of the redox reactions that GSH participates in (Prabhu et al., 2004; Sharma et al., 2004; Torres-Rivera and Landa, 2008; Yang et al., 2002a). By functioning as Se-independent GPXs, these GSTs provide protection against lipid peroxidation by terminating lipid peroxidation cascade through the reduction of fatty acid hydroperoxides (FA-OOH), and phospholipid hydroperoxides (PL-OOH) (Hayes et al., 2005; Sharma et al., 2004; Yang et al., 2001; 2002b). Since these GST enzymes (e.g., hGSTA1-1 and hGSTA2-2) can use membrane PL-OOH, such as 1-palmitoyl-2-(13-hydroperoxy-cis-9,trans-11-octadecadienoyl)-L-3-phosphatidylcholine and phosphatidylcholine hydroperoxide, as substrates in situ, they can protect cell membranes at the site of damage (Hayes et al., 2005; Prabhu et al., 2004; Sharma et al., 2004; Yang et al., 2001; 2002a). Furthermore, as oxidative stress and subsequent lipid peroxidation are the common secondary effects of many drugs and xenobiotics, GSTs with Se-independent peroxidases play a crucial role in protection against xenobiotics by reducing lipid hydroperoxides. For example, GSTs with GSH peroxidase activity are implicated in pyrethroid resistance in insects (Vontas et al., 2001; 2002) and herbicide resistance in plants (Edwards et al., 2000).

Some GSTs can effectively detoxify the organochlorine insecticide DDT by functioning as GSH-dependent DDTase. A number of insect GSTs have been shown to catalyze the dehydrochlorination of the DDT to the noninsecticidal metabolite DDE by using GSH as a cofactor rather than as a substrate (Chen et al., 2003; Li et al., 2007; Lumjuan et al., 2005; Ortelli et al., 2003; Ranson et al., 2001). Besides catalyzing conjugation, reduction and dehydrochlorination reactions, GSTs also contribute to intracellular and circulatory transport, sequestration, and disposition of endogenous lipophilic compounds and xenobiotics by acting as binding and transport proteins of hydrophobic non-substrates (Hayes et al., 2005; Li et al., 2007). This non-enzymatic binding activity can reduce the number of toxic compounds that reaches their target sites (Kostaropoulos et al., 2001) and speed up their elimination.

Regulation of GSTs Expression

GST enzymes are expressed at a basal level in normal conditions. Upon exposure to low levels of oxidative stress or a diverse range of endogenous and exogenous harmful electrophiles, expression of GST enzymes is strongly induced (Figure 2) (Blanchette et al., 2007; Cheng et al., 2001; Dinkova-Kostova et al., 2001; Surh et al., 2008).

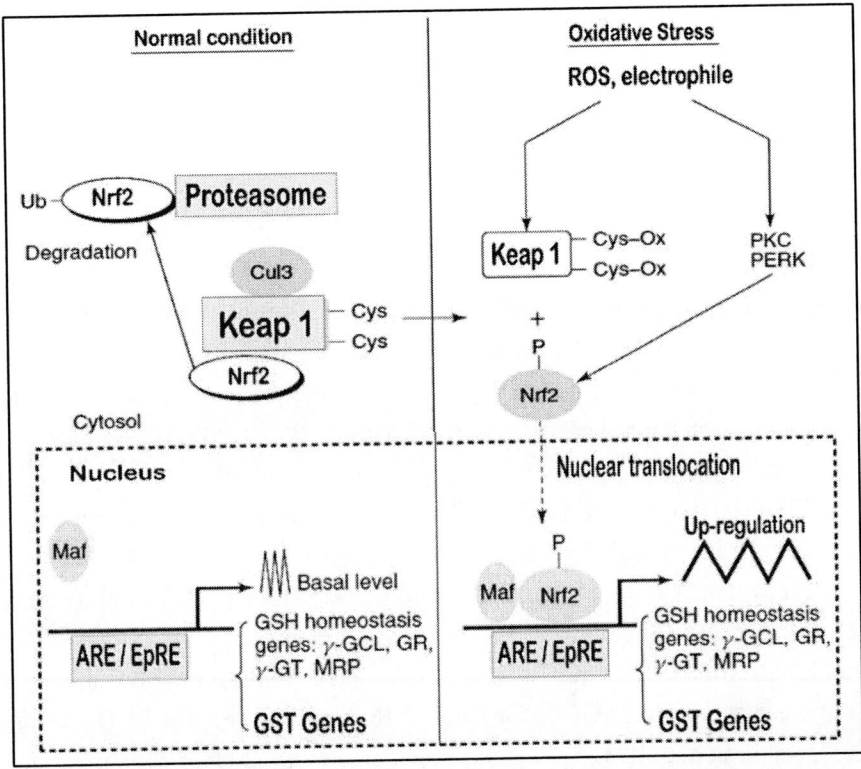

Figure 2. Redox and electrophilic regulation of GSH homeostasis and GST expression. Under normal conditions, Nrf2 is captured in the cytoplasm by the Keap1-Cul3 repressor complex for degradation by proteasome, and GSH homeostasis and GST genes are expressed at basal levels. Upon oxidative stress or electrophilic stimuli, Nrf2 is released from the Keap1-Cul3 repressor complex due to the oxidation/modification of the critical cysteine thiols of Keap1 or phosphorylation of Nrf2 by kinases such as PKC or PERK. The free Nrf2 is translocated to the nucleus, where it heterodimerizes with Maf proteins and then binds to the antioxidant responsive element (ARE) or electrophile responsive element (EpRE) in the 5' promoter region of GSH homeostasis and GST genes. As a result, transcriptions of GSH homeostasis and GST genes are coordinately up-regulated (induced), leading to a concomitant increase in cellular GSH levels and GST enzymes. Grey, Active forms of the molecules. γ-GCL, glutamylcysteine ligase; GR, glutathione reductase; γ-GT, γ-glutamyl transpeptidase; MRP, multidrug-resistance protein transporter; GST, glutathione-S-transferase (Adapted from Li, 2011).

The electrophilic compounds that induce expression of GST genes are often the hydrophobic substrates of GST enzymes, which include lipid peroxidation products (4-hydroxynonenal (4-HNE)), chemoprotective agents (oltipraz, sulforaphane, 3H-1,2-dithiole-3-thione), phytochemicals (cruciferous sulforaphane, furanocoumarins, indoles, flavonoids, α,β-unsaturated carbonyls and glucosinolates) and various toxicants or pro-oxidants(t-butylhydroquinone, butylhydroxyanisole, thiazoles) (Kim and Lee, 2007; Li et al., 2007; Surh et al., 2008; Zhang et al., 2006; Zhu et al., 2008b). These inducers belong to nine distinct chemical classes, including oxidizable diphenols and quinones, Michael reaction acceptors (olefins or acetylenes conjugated to electron-withdrawing groups), isothiocyanates, hydroperoxides, trivalent arsenic derivatives, divalent heavy metal cations (Hg^{2+}, Cd^{2+}), vicinal dithiols, 1,2-dithiole-3-thiones, and carotenoids and other conjugated polyenes (Dinkova-Kostova et al., 2001). Nonetheless, they all can covalently modify sulfhydryl groups by alkylation, oxidation or reduction, which is responsible for the initial 'sensing' of these inducers by the actin-binding protein repressor Keap1, which contains several critical cysteines (Cys-151, Cys-273 and Cys-288) for modification (Figure 2) (Dinkova-Kostova et al., 2001; Hayes et al., 2005). Interestingly, induction of GST genes is often paralleled with induction of GSH homeostasis genes (γ-GCL, GS, GR, MRP, γ-GT) (Figure 2) (Li et al., 2005; Shih et al., 2003; Surh et al., 2008; Zhu et al., 2008a; 2008b; 2008c). Such a coordinated induction of GSH homeostasis genes and GST genes (also other Phase 2 detoxification genes) leads to a concomitant increase in cellular GSH levels and GST enzymes, which enables the cell to survive exposure to harmful xenobiotics and oxidative stress (Cao et al., 2003; Cheng et al., 2001; Dinkova-Kostova et al., 2001; Li et al., 2005). Thus, it is not a surprise that GST genes, like GSH homeostasis genes, also contain ARE/EpRE (consensus sequence TGACNNNGC) in their 5'-promoter regions and are regulated by the very same Nrf2-Keap1 signaling pathway (Figure 2) (Hayes et al., 2005; Kim and Lee, 2007; Surh et al., 2008).

Association of Glutathione-S-Transferases with Bladder Cancer

Bladder cancer is the 9th most common cancer and a leading cause of cancer-related death worldwide. An estimated 429,800 new cases of bladder cancer and 165,100 deaths occurred in 2012 worldwide (Torres et al., 2015), as well as 550,000 new cases and 199,922 deaths in 2018 and these numbers are expected to double in the upcoming years (Bray et al., 2018). The disease is highly recurring and do frequently progress to a muscle invasive phenotype which necessitate a vigilant and continuous monitoring protocol (Sanli et al., 2017). Despite advances in diagnostic and treatment modalities, bladder cancer remains source of co-morbidity and continues to pose challenges for clinicians given that patients' outcome being solely dependent on the grading and staging system (Youssef and Lotan, 2011). Therefore, a deeper understanding of the bladder cancer pathogenesis and associated mechanisms will undoubtedly improve patients' outcome via prevention of disease progression and recurrence.

It is well documented that occupational exposure to chemical carcinogens including aromatic amines and polycyclic aromatic hydrocarbons is associated with the risk of bladder cancer development (Boada et al., 2015; Sanli et al., 2017). Kellen et al. (2007) reported an increased risk of developing bladder cancer associated with cumulative exposure to aromatic amines, but not to PAHs and diesel. In an independent study, Ferrís et al. (2013) concluded that bladder cancer is a result of the interaction between constitutional and environmental risk factors including aromatic amines and polycyclic aromatic hydrocarbons. The involvement of environmental factors such as cigarette smoking in bladder carcinogenesis has been extensively investigated (Brennan et al., 2000; Saint-Jacques et al., 2018). Recent evidence supports the dynamic interplay between environmental factors and other co-factors, including genetic predisposition, in the pathogenesis of bladder cancer (Glaser et al., 2017).

Protecting against carcinogen-induced and chemotherapy induced oxidative stress involves a series of event characterized by the activation of phase-II cellular detoxifying enzymes; Glutathione S-transferases (GSTs) or N-acetyltransferases (NATs) (Guengerich, 2000). GSTs enzymes superfamily consist of at least 16 genes located on more than 7 chromosomes (Strange et al., 2001). Although they are structurally different with distinct evolutionary origins, all GSTs isoenzymes are functionally similar in protection against electrophiles and oxidative stressors. GSTs play a critical protective anticancer role through glutathione conjugation with a range of potentially cytotoxic exogenous or endogenous molecules making them less toxic. Besides, GSTs are able to regulate the induction of other proteins and enzymes which is important for cellular functions. The polymorphisms affect the enzyme activity, leading to increased genotoxic damage and affect the transportation of steroid hormones, causing the development of cancer eventually (Albarakati et al., 2019; McIlwain et al., 2006). GSTs are essential for maintaining genomic integrity because electrophilic compounds could damage the DNA (Huang et al., 2015). Allelic polymorphisms in these genes elicit changes in enzyme activities leading to biotransformation and play important role in the development and progression of different cancers, such as lung, colorectal, leukemia, breast and bladder cancers.

It has long been perceived that bladder cancer is a result of occupational and environmental exposure to carcinogens and tobacco smoking, however, the exact mechanisms of bladder carcinogenesis remain unclear. Recent findings suggested that genetic factors contribute potentially, through mutations in key genes, in the etiology and pathogenesis of bladder cancer (Brennan et al., 2000; Knowles and Hurst, 2015; Saint-Jacques et al., 2018). Glutathione S-Transferases, are members of a large gene family of cytosolic phase II xenobiotic metabolizing enzymes involved in catalyzing and detoxifying a variety of carcinogens including reactive electrophilic compounds (Stranges et al., 2001). Members of the GST family play an important role in cellular defense through conjugation of xenobiotics with sulfhydryl group and promoting their excretion at later stage (Lang and Pelkonen, 1999; Stranges et al., 2001). It has been proposed that polymorphisms in members of GST of carcinogen detoxifying gene family as well as in NAT2 confer increased risk of bladder cancer (Knowles and Hurst, 2015). Moreover, increased expression of GST family members, especially GSTM1, GSTP1 and GSTM1, was reported in several human solid tumors and is believed to confer resistance to various platinum base chemotherapy drugs and metformin through regulation of many genes and molecular pathways

(Allocati et al., 2018; Sawers et al., 2014). A recent investigation conducted by Savik-Radojevic et al. (2103) has evidenced the presence of significant amount of 8-hydroxy-2′-deoxyguanosine (8-OHdG), in the urine of bladder cancer patients indicating oxidative DNA damage. They have also reported high level of 8-OHdG in patients having null GST M1 genotypes (Savik-Radojevic et al., 2103). Mechanistically, it is believed that polymorphisms in genes involved in drug-metabolizing enzymes may result in drastic changes in carcinogens biotransformation leading to increased cancer susceptibility (Sanli et al., 2017).

Over the last 2 decades, plentiful studies have been carried out to investigate the association between GSTs and the risk of bladder cancer, but these studies have reported conflicting results. A single study might fail to demonstrate the complicated genetic relationship due to small sample size, but meta-analysis and review studies could increase the statistical power through detecting overall effects. Previously, meta-analysis has been performed to find out the relationship between GSTM1, GSTT1, GSTP1, and bladder cancer, respectively (Gong et al., 2012; Jiang et al., 2011; Wang et al., 2013; Wu et al., 2013; Zhang et al., 2011;). In this chapter, we have included results of different meta-analysis and review studies to conclude the overall impacts of GSTM1, GSTT1 and GSTP1 in the susceptibility of bladder cancer.

GSTM1 and Bladder Cancer

GSTM1 gene is located on chromosome 1p13.3, and the homozygous deletion (GSTM1 null genotype) is the most common polymorphic variant of GSTM1 gene which is characterized by abolished enzyme activity (Lin et al., 1994). The relationship between the genetic polymorphism of GSTM1 and the risk of bladder cancer have been studied by several investigators and reported controversial results among different populations. Previous epidemiological studies showed an association between the homozygous deletion of GSTM1 and increased risk of bladder, lung, and colorectal cancers (Ates et al., 2005; Benhamou et al.,

2002; Singh et al., 2008). However other studies failed to establish the association between GSTM1 null and the risk of several types of cancers (Gorukmez et al., 2016; Piao et al., 2009; Yin et al., 2017). A meta-analysis including 72 studies, which contained 20,239 cases and 24,393 controls, assessed the relationship between the GSTM1-null genotype and bladder cancer susceptibility (Zhou et al., 2018). Average distribution frequency of the GSTM1-null genotype was 56.15% in the bladder cancer group and 46.97% in the control group, indicating that the GSTM1-null genotype was associated with bladder cancer risk in the overall population, and individually in whites, Africans and Asians (overall population: OR = 1.40, 95% CI: 1.31–1.48, $P < 0.001$; whites: OR = 1.39, 95% CI: 1.26–1.54, $P < 0.001$; Africans: OR = 1.54, 95% CI: 1.16–2.05, $P = 0.003$; Asians: OR = 1.45, 95% CI: 1.33–1.59, $P < 0.001$); as well as in controls from both hospital-based and population-based studies that included both high- and low-quality studies (Zhou et al., 2018).

Another study comprising 48 studies including 11,473 cases and 13,795 controls were reported (Yu et al., 2016). Overall, significant associations between individuals who carried GSTM1 null genotype and increased bladder cancer risk were observed (OR = 1.39, 95%CI 1.28–1.51) (Yu et al., 2016). When stratified by ethnicity, significant difference was detected in Caucasian (OR = 1.39, 95% CI 1.23–1.58) and Asian populations (OR = 1.45, 95% CI 1.31–1.61) instead of African (OR = 1.23, 95% CI 0.95–1.59) or Mixed populations (OR = 1.16, 95% CI 0.93–1.45) (Yu et al., 2016). In addition, a total of nine studies with 1646 bladder cancer cases and 1938 controls were included to systematically explore the association between GSTM1 and GSTT1 polymorphisms and bladder cancer risk (Chen et al., 2018). From the combined statistical results, we found a significant association between GSTM1-null genotype with the risk of bladder cancer in the overall analysis (Chen et al., 2018).

Furthermore, 46 studies described the relationship between GSTM1 polymorphism and bladder cancer susceptibility, involving 28270 individuals (Yu et al., 2017). The pooled meta-analysis showed that the GSTM1 null genotype was associated with increased risk of bladder cancer. The pooled summary of the OR was 1.36 (95% CI: 1.25-1.47, p< 0.01) (Yu et al., 2017). The GSTM1 null genotype was associated with the elevated risk of bladder cancer in Caucasians (OR = 1.34, 95% CI = 1.21-1.48) and Asians (OR = 1.50, 95% CI = 1.31-1.71) (Yu et al., 2017). Stratified analyses of population-based association showed a significant association of elevated bladder cancer risk with GSTM1 deletion in hospital-based (HB) studies (OR = 1. 42, 95% CI = 1.30-1.56) and population-based (PB) studies (OR = 1.22, 95% CI = 1.07-1.40) (Yu et al., 2017).

GSTT1 and Bladder Cancer

GSTT1 gene is located on chromosome 22q11.23, and the most common polymorphic variant of GSTT1 gene is the homozygous deletion (GSTT1 null genotype) (Yu et al., 2016). People with the GSTT1 null genotype was reported to have decreased enzyme activity and decreased ability to detoxify the environmental and dietary agents, especially 1, 3-butadiene and ethylene oxide, which could induce chromosomal damage and make people more susceptible to cancer (Harris et al., 1998). Many studies have investigated the relationship between the genetic polymorphism of GSTT1 and the risk of bladder cancer, but the association remains controversial among different populations (Yu et al., 2016; Zhou et al., 2018).

A meta-analysis including 61 studies, which contained 13,041 bladder cancer cases and 16,739 controls, were analyzed to assess the relationship between the GSTT1-null genotype and bladder cancer susceptibility (Zhou et al., 2018). The GSTT1-null genotypic distribution was 29.58% in the bladder cancer group and 26.67% in the control group, indicating that the GSTT1-null genotype was higher in the bladder cancer cases compared with the controls (Zhou et al., 2018). In this study, they reported that the GSTT1-null genotype was associated with bladder cancer risk in the overall population, and controls from hospital-based studies that included high-quality studies; but not with bladder cancer risk in Whites, Africans, Asians or controls from population-based studies that included low-quality studies (overall population: OR = 1.11, 95% CI: 1.01–1.22, p = 0.03; whites: OR = 1.16, 95% CI: 0.99–1.36, p = 0.07; Africans: OR = 1.07, 95% CI: 0.65– 1.76, p = 0.79; Asians: OR = 1.05, 95% CI: 0.91–1.22, p = 0.51) (Zhou et al., 2018). However, 57 studies including 12,369 bladder cancer cases and 15,333 control subjects were analyzed and depicted no significant association between GSTT1 polymorphism and bladder cancer susceptibility (OR = 1.11, 95% CI: 1.00–1.22) (Yu et al., 2016). In addition, the subgroup analysis by ethnicity, showed significant associations between GSTT1 null genotype and bladder cancer risk only in Caucasians (OR = 1.25, 95% CI: 1.09–1.44) (Yu et al., 2016).

Another investigation comprising 54 studies described the relationship between GSTT1 polymorphism and bladder cancer susceptibility, involving 26622 individuals (Yu et al., 2017). Further, a pooled meta-analysis showed that the GSTT1 null genotype was associated with elevated risk of bladder cancer and the pooled summary of the OR was 1.13 (95% CI: 1.02-1.25, p < 0.01) (Yu et al., 2017). Subgroup analyses were performed on the different ethnicity, population based and smoking. The results suggested that the GSTT1 null genotype was associated with the elevated risk of bladder cancer in Caucasians (OR = 1.23, 95% CI = 1.08-1.40) (Yu et al., 2017). However, no significant association was found in Asians, Africans, and multiracial subjects. Stratified analyses of population-based association showed a weak association of elevated bladder cancer risk with GSTT1 deletion in HB studies (OR = 1.11, 95% CI = 0.98-1.27) and PB studies (OR = 1.15, 95% CI = 1.00-1.33), but without statistical significance (Yu et al., 2017).

GSTP1 and Bladder Cancer

GSTP1 is encoded by a single gene located on chromosome 11 (Saint-Ruf et al., 1991). The common functional GSTP1 polymorphism at codon 105 is an A to G substitution resulting in an amino acid switch from isoleucine to valine (Ile105Val) and lowering the catalytic activity of GSTP1 enzyme (Ali-Osman et al., 1997). The decreased detoxification capacity of the GSTP1 enzyme resulted in differences in chemotherapeutic responses. The increased expression of the GSTP1 Val105 genotype was shown to be associated with a variety of tumors, such as ovarian, breast, colon, lymphoma, and pancreas (Tew et al., 2011). The hypothesis that GSTP1 variants modulate the risk of urinary bladder cancer has also been investigated (Tew et al., 2011; Zhang et al., 2016). However, inconclusive results have been reported on the association between GSTP1 gene polymorphisms and the risk of bladder cancer while a number of studies identified an obvious association between GSTP1 polymorphisms Ile105Val and bladder carcinoma risk (Fontana et al., 2009; Kellen et al., 2007; Safarinejad et al., 2013), other studies illustrated that there are no association between GSTP1 Ile105Val polymorphism and bladder cancer (Matic et al., 2013; Yu et al., 2016).

A meta-analysis involving 23 studies including 5080 bladder cancer cases and 6187 controls was conducted which revealed no significant association between GSTP1 Ile105Val polymorphism and bladder cancer risk (OR = 1.07, 95% CI 0.96–1.20) (Yu et al., 2016). No significant relationship was observed between GSTP1 polymorphism and bladder cancer risk in patients when stratified by ethnicity (Yu et al., 2016). Meanwhile, there seems no relationship between GSTP1 polymorphism and the susceptibility of bladder cancer when stratified by source of controls (Yu et al., 2016).

CONCLUSION

GSH and GSTs are two primary lines of defense that protect human and other organisms from both acute and chronic toxicities of electrophiles and reactive oxygen/nitrogen species. In addition to their well-established GSH-conjugating enzymatic activity, GSTs of the M, T, and P classes have been shown to modulate signaling pathways that control cell proliferation, cell differentiation, and cell death by interacting with important signaling proteins in a non-enzymatic way. Genetic polymorphisms of human GST enzymes (GSTM1, GSTT1, and GSTP1) have said to be associated with increased cancer risks. In this chapter we have analyzed several meta-analysis and review papers and found that the GSTM1-null, and GSTT1-null genotypes might be associated with the onset of bladder cancers. On the other hand, no significant associations were observed between GSTP1 polymorphism and bladder cancer susceptibility. As specific GST isozymes are overexpressed in a wide variety of tumors resistant to drugs, GSTs have emerged as a promising therapeutic target for the development of novel cancer drugs. However, taking the restriction of sample size into consideration, analysis with larger and well-designed studies is required to validate that findings.

REFERENCES

Ahmad, S. (1992). Biochemical defence of pro-oxidant plant allelochemicals by herbivorous insects. *Biochemical Systematics and Ecology*, 20:269–96.

Albarakati, N., Khayyat, D., Dallol, A., Al-Maghrabi, J. and Nedjadi, T. (2019). The prognostic impact of GSTM1/GSTP1 genetic variants in bladder Cancer. *BMC Cancer*, 19(1):991.

Ali-Osman, F., Akande, O., Antoun, G., Mao, J. X. and Buolamwini, J. (1997). Molecular cloning, characterization, and expression in Escherichia coli of full-length cDNAs of three human glutathione S-transferase pi gene variants. Evidence for differential catalytic activity of the encoded proteins. *Journal of Biological Chemistry*, 272(15):10004–12.

Allocati, N., Masulli, M., Di-Ilio, C. and Federici, L. (2018). Glutathione transferases: substrates, inhibitors and pro-drugs in cancer and neurodegenerative diseases. *Oncogenesis*, 7(1):8.

Apel, K. and Hirt, H. (2004). Reactive oxygen species: metabolism oxidative stress, and signal transduction. *Annual Review of Plant Biology*, 55:373–99.

Ates, N. A., Tamer, L., Ates, C., Ercan, B., Elipek, T., Ocal, K. and Camdeviren, H. (2005). Glutathione S-transferase M1, T1, P1 genotypes and risk for development of colorectal cancer. *Biochemical Genetics*, 43(3–4):149–63.

Benhamou, S., Bonastre, J., Groussard, K., Radvanyi, F., Allory, Y. and Lebret, T. (2016). A prospective multicenter study on bladder cancer: the COBLAnCE cohort. *BMC Cancer*, 16:837.

Benhamou, S., Lee, W. J., Alexandrie, A. K., Boffetta, P., Bouchardy, C., Butkiewicz, D., Brockmoller, J., Clapper, M. L., Daly, A., Dolzan, V., Ford, J., Gaspari, L., Haugen, A., Hirvonen, A., Husgafvel-Pursiainen, K., Ingelman-Sundberg, M., Kalina, I., Kihara, M., Kremers, P., Le Marchand, L., London, S. J., Nazar-Stewart, V., Onon-Kihara, M., Rannug, A., Romkes, M., Ryberg, D., Seidegård, J., Shields, P., Strange, R. C., Stucker, I., To-Figueras, J., Brennan, P. and Taioli, E. (2002). Meta- and pooled analyses of the effects of glutathione S-transferase M1 polymorphisms and smoking on lung cancer risk. *Carcinogenesis*, 23(8):1343–50.

Blair, I. A. (2006). Endogenous glutathione adducts. *Current Drug Metabolism*, 7:853–72.

Blanchette, B., Feng, X. and Singh, B. R. (2007). Marine glutathione S-transferases. *Marine Biotechnology*, 9:513–42.

Boada, L. D., Henríquez-Hernández, L. A., Navarro, P., Zumbado, M., Almeida-González, M., Camacho, M., Álvarez-León, E. E., Valencia-Santana, J. A. and Luzardo, O. P. (2015). Exposure to polycyclic aromatic hydrocarbons (PAHs) and bladder cancer: evaluation from a gene environment perspective in a hospital-based case-control study in the Canary Islands (Spain). *International Journal for Occupational and Environmental Health*, 21(1):23–30.

Bray, F., Ferlay, J., Soerjomataram, I., Siegel, R. L., Torre, L. A. and Jemal, A. (2018). Global cancer statistics 2018: GLOBOCAN estimates of incidence and mortality worldwide for 36 cancers in 185 countries. *CA: A Cancer Journal for Clinicians*, 68(6):394–424.

Brennan, P., Bogillot, O., Cordier, S., Greiser, E., Schill, W., Vineis, P., Lopez-Abente, G., Tzonou, A., Chang-Claude, J., Bolm-Audorff, U., Jöckel, K. H., Donato, F., Serra, C., Wahrendorf, J., Hours, M., T'Mannetje, A., Kogevinas, M. and Boffetta, P. (2000). Cigarette smoking and bladder cancer in men: a pooled analysis of 11 case-control studies. *International Journal of Cancer*, 86(2):289–94.

Cao, Z., Hardej, D., Trombetta, L. D. and Li, Y. (2003). The role of chemically induced glutathione and glutathione S-transferase in protecting against 4-hydroxy-2-nonenal-mediated cytotoxicity in vascular smooth muscle cells. *Cardiovascular Toxicology*, 3:165–77.

Chen, C. H., Chiou, H. Y., Hsueh, Y. M., Chen, C. J., Yu, H. J. and Pu, Y. S. (2009). Clinicopathological characteristics and survival outcome of arsenic related bladder cancer in Taiwan. *Journal of Urology*, 181:547–52.

Chen, C. H., Shun, C. T., Huang, K. H., Huang, C. Y., Yu, H. J. and Pu, Y. S. (2008). Characteristics of female non-muscle-invasive bladder cancer in Taiwan: association with upper tract urothelial carcinoma and end-stage renal disease. *Urology*, 71:1155–60.

Chen, D. K., Huang, W. W., Li, L. J., Pan, Q. W. and Bao, W. S. (2018). Glutathione S-transferase M1 and T1 null genotypes and bladder cancer risk: A meta-analysis in a single ethnic group. *Journal of Cancer Research and Therapy*, 14: S993-7.

Chen, L., Hall, P. R., Zhou, X. E., Ranson, H., Hemingway, J. and Meehan, E. J. (2003). Structure of an insect delta class glutathione S-transferase from a DDT-resistant strain of the malaria vector *Anopheles gambiae*. *Acta Crystallographica - Section D: Biological Crystallography*, 59:2211–7.

Cheng, J. Z., Sharma, R., Yang, Y., Singhal, S. S., Sharma, A., Saini, M. K., Singh, S. V., Zimniak, P., Awasthi, S. and Awasthi, Y. C. (2001). Accelerated metabolism and exclusion of 4-hydroxynonenal through induction of RLip76 and hGST5.8 is an early adaptive response of cells to heat and oxidative stress. *Journal of Biological Chemistry*, 276:41213–23.

Dinkova-Kostova, A. T., Massiah, M. A., Bozak, R. E., Hicks, R. J. and Talalay, P. (2001). Potency of Michael reaction acceptors as inducers of enzymes that protect against carcinogenesis depends on their reactivity with sulfhydryl groups. *Proceedings of the National Academy of Sciences*, 98:3404–9.

Edwards, R., Dixon, P. D. and Walbot, V. (2000). Plant glutathione S-tranferases: enzymes with multiple functions in sickness and health. *Trends in Plant Science*, 5:193–8.

Ferrís, J., Garcia, J., Berbel, O. and Ortega, J. A. (2013). Constitutional and occupational risk factors associated with bladder cancer. *Actas Urologicas Espanola*, 37(8):513–22.

Fontana, L., Delort, L., Joumard, L., Rabiau, N., Bosviel, R., Satih, S., Guy, L., Boiteux, J-P., Bignon Y-J., Chamoux, A. and Bernard-gallon, D. J. (2009). Genetic polymorphisms in CYP1A1, CYP1B1, COMT, GSTP1 and NAT2 genes and association with bladder cancer risk in a French cohort. *Anticancer Research*, 29(5):1631–5.

Forman, H. J., Torres, M. and Fukuto, J. (2002). Redox signaling. *Molecular and Cellular Biochemistry*, 234–235:49–62.

Franco, R., Schoneveld, O. J., Pappa, A. and Panayiotidis, M. I. (2007). The central role of glutathione in the pathophysiology of human diseases. *Archives of Physiology and Biochemistry*, 113:234–58.

Glaser, A. P., Fantini, D., Shilatifard, A., Schaeffer, E. M. and Meeks, J. J. (2017). The evolving genomic landscape of urothelial carcinoma. *Nature Reviews Urology*, 14:215.

Gong, M., Dong, W. and An, R. (2012). Glutathione S-transferase T1 polymorphism contributes to bladder cancer risk: a meta-analysis involving 50 studies. *DNA Cell Biology*, 31:1187–97.

Gorukmez, O., Yakut, T., Gorukmez, O., Sag, S.O., Topak, A., Sahinturk, S. and Kanat, O. (2016). Glutathione S-transferase T1, M1 and P1 genetic polymorphisms and susceptibility to colorectal cancer in Turkey. *Asian Pacific Journal of Cancer Prevention*, 17(8):3855–9.

Guengerich, F. P. (2000). Metabolism of chemical carcinogens. *Carcinogenesis*, 21(3):345–51.

Harris, M. J., Coggan, M., Langton, L., Wilson, S. R. and Board, P. G. (1998). Polymorphism of the Pi class glutathione S-transferase in normal populations and cancer patients. *Pharmacogenetics*, 8:27–31.

Hayes, J. D., Flanagan, J. U. and Jowsey, I. R. (2005). Glutathione transferases. *Annual Review of Pharmacology and Toxicology*, 45:51–88.

Huang, W., Shi, H., Hou, Q., Mo, Z. and Xie, X. (2015). GSTM1 and GSTT1 polymorphisms contribute to renal cell carcinoma risk: evidence from an updated meta-analysis. *Scientific Reports*, 5:17971.

Jiang, Z., Li, C. and Wang, X. (2011). Glutathione S-transferase M1 polymorphism and bladder cancer risk: a meta-analysis involving 33 studies. *Experimental Biology and Medicine*, 236:723–8.

John, B. A. and Said, N. (2017). Insights from animal models of bladder cancer: recent advances, challenges, and opportunities. *Oncotarget*, 8:57766–81.

Kellen, E., Hemelt, M., Broberg, K., Golka, K., Kristensen, V. N., Hung, R. J., Matullo, G., Mittal, R. D., Porru, S., Povey, A., Schulz, W. A., Shen, J., Buntinx, F., Zeegers, M. P. and Taioli, E. (2007). Pooled analysis and meta-analysis of the glutathione S-transferase P1 Ile 105Val polymorphism and bladder cancer: a HuGE-GSEC review. *American Journal of Epidemiology*, 165(11):1221–30.

Kellen, E., Zeegers, M., Paulussen, A., Vlietinck, R., Vlem, E. V., Veulemans, H. and Buntinx, F. (2007). Does occupational exposure to PAHs, diesel and aromatic amines interact with smoking and metabolic genetic polymorphisms to increase the risk on bladder cancer? The Belgian case control study on bladder cancer risk. *Cancer Letter*, 245(1–2):51–60.

Kim, S. G. and Lee, S. J. (2007). PI3K, RSK, and mTOR signal networks for the GST gene regulation. *Toxicological Sciences*, 96:206–13.

Knowles, M. A. and Hurst, C. D. (2015). Molecular biology of bladder cancer: new insights into pathogenesis and clinical diversity. *Nature Reviews Cancer*, 15(1):25–41.

Kostaropoulos, I., Papadopoulos, A. I., Metaxakis, A., Boukouvala, E. and Papadopoulou-Mourkidou, E. (2001). Glutathione S-transferase in the defence against pyrethroids in insects. *Insect Biochemistry and Molecular Biology*, 31:313–9.

Lang, M. and Pelkonen, O. (1999). Metabolism of xenobiotics and chemical carcinogenesis. *IARC Scientific Publications*, 148:13–22.

Li, X. (2011). Glutathione and Glutathione-S-Transferase in Detoxification Mechanisms. In *General, Applied and Systems Toxicology,* John Wiley and Sons, 2011. DOI: 10.1002/9780470744307.gat166.

Li, X., Schuler, M. A. and Berenbaum, M. R. (2007). Molecular mechanisms of metabolic resistance to synthetic and natural xenobiotics. *Annual Review of Entomology*, 52:231–53.

Li, Y., Cao, Z., Zhu, H. and Trush, M. A. (2005). Differential roles of 3H-1,2-dithiole-3-thione-induced glutathione, glutathione S-transferase and aldose reductase in protecting against 4-hydroxy-2-nonenal toxicity in cultured cardiomyocytes. *Archives of Biochemistry and Biophysics*, 439:80–90.

Lin, H. J., Han, C-Y., Bernstein, D. A., Hsiao, W., Lin, B. K. and Hardy, S. (1994). Ethnic distribution of the glutathione transferase mu 1-1 (GSTM1) null genotype in 1473 individuals and application to bladder cancer susceptibility. *Carcinogenesis*, 15(5):1077–81.

Łukasik, I. and Goławska, S. (2007). Activity of Se-independent glutathione peroxidase and glutathione reductase within cereal aphid tissues. *Biological Letters*, 44:31–39.

Lumjuan, N., McCarroll, L., Prapanthadara, L., Hemingway, J. and Ranson, H. (2005). Elevated activity of an Epsilon class glutathione transferase confers DDT resistance in the dengue vector, Aedes aegypti. *Insect Biochemistry and Molecular Biology*, 35:861–71.

Mates, J. M. (2000). Effects of antioxidant enzymes in the molecular control of reactive oxygen species toxicology. *Toxicology*, 153:83–104.

Matic, M., Pekmezovic, T., Djukic, T., Mimic-Oka, J., Dragicevic, D., Krivic, B., Suvakov, S., Savic-Radojevic, A., Pljesa-Ercegovac, M., Tulic, C., Coric, V. and Simic, T. (2013). GSTA1, GSTM1, GSTP1, and GSTT1 polymorphisms and susceptibility to smoking-related bladder cancer: a case-control study. *Urology Oncology*, 31(7):1184–92.

McIlwain, C. C., Townsend, D. M. and Tew, K. D. (2006). Glutathione Stransferase polymorphisms: cancer incidence and therapy. *Oncogene*, 25:1639–48.

Monks, T. J., Anders, M. W., Dekant, W., Stevens, J. L., Lau, S. S. and Van-Bladeren, P. J. (1990). Glutathione conjugate mediated toxicities. *Toxicology and Applied Pharmacology*, 106:1–19.

Oliveira, P. A., Vasconcelos-Nobrega, C., Gil da Costa, R. M. and Arantes-Rodrigues, R. (2018). The N-butyl-N-4-hydroxybutyl nitrosamine mouse urinary bladder Cancer model. *Methods in Molecular Biology*, 1655:155–67.

Ortelli, F., Rossiter, L. C., Vontas, J., Ranson, H. and Hemingway, J. (2003). Heterologous expression of four glutathione transferase genes genetically linked to a major insecticide-resistance locus from the malaria vector *Anopheles gambiae*. *Biochemical Journal*, 373:957–63.

Oztetik, E. (2008). A tale of plant glutathione S-transferases: since 1970. *The Botanical Review*, 74:419–37.

Piao, J-M., Shin, M-H., Kweon, S-S., Kim, H. N., Choi, J-S., Bae, W-K., Shim, H-J., Kim, H-R., Park, Y-K., Choi, Y-D and Kim, S-H. (2009). Glutathione-S-transferase (GSTM1, GSTT1) and the risk of gastrointestinal cancer in a Korean population. *World Journal of Gastroenterology*, 15(45):5716–21.

Prabhu, K. S., Reddy, P. V., Jones, E. C., Liken, A. D. and Reddy, C. C. (2004). Characterization of a class alpha glutathione-S-transferase with glutathione peroxidase activity in human liver microsomes. *Archives of Biochemistry and Biophysics*, 424:72–80.

Ranson, H. and Hemingway, J. (2005). Glutathione transferases. In Gilbert, L. I., Iatrou, K. and Gill, S. S. (Eds), *Comprehensive Molecular Insect Science, Pharmacology*, Vol. 5. Elsevier, Oxford, pp. 383–402.

Ranson, H., Rossiter, L., Ortelli, F., Jesen, B., Wang, X., Roth, C. W., Collins, F. H. and Hemingway, J. (2001). Identification of a novel class of insect glutathione S-transferases involved in resistance to DDT in the malaria vector *Anopheles gambiae*. *Biochemical Journal*, 359:295–304.

Reed, D. J. (1995). Toxicity of oxygen. In De Matteis, F. and Smith, L. L. (Eds), *Molecular and Cellular Mechanisms of Toxicity*. 1st Edition, CRC Press, Florida, pp. 35–68.

Reszka, E. (2012). Selenoproteins in bladder cancer. *Clinica Chimica Acta*, 413:847–54.

Safarinejad, M. R., Safarinejad, S., Shafiei, N. and Safarinejad, S. (2013). Association of genetic polymorphism of glutathione S-transferase (GSTM1, GSTT1, GSTP1) with bladder cancer susceptibility. *Urology Oncology*, 31(7):1193–203.

Saint-Jacques, N., Brown, P., Nauta, L., Boxall, J., Parker, L. and Dummer, T. J. B. (2018). Estimating the risk of bladder and kidney cancer from exposure to low-levels of arsenic in drinking water, Nova Scotia, Canada. *Environment International*, 110:95–104.

Saint-Ruf, C., Malfoy, B., Scholl, S., Zafrani, B. and Dutrillaux, B. (1991). GST pi gene is frequently coamplified with INT2 and HSTF1 proto-oncogenes in human breast cancers. *Oncogene*, 6(3):403–6.

Sanli, O., Dobruch, J., Knowles, M. A., Burger, M., Alemozaffar, M., Nielsen, M. E. and Lotan, Y. (2017). Bladder cancer. *Nature Reviews Disease Primers*, 3:17022.

Savic-Radojevic, A., Djukic, T., Simic, T., Pljesa-Ercegovac, M., Dragicevic, D., Pekmezovic, T., Cekerevac, M., Santric, V. and Matic, M. (2013). GSTM1-null and GSTA1-low activity genotypes are associated with enhanced oxidative damage in bladder cancer. *Redox Report*, 18(1):1-7.

Sawers, L., Ferguson, M. J., Ihrig, B. R., Young, H. C., Chakravarty, P., Wolf, C. R. and Smith, G. (2014). Glutathione S-transferase P1 (GSTP1) directly influences platinum drug chemosensitivity in ovarian tumour cell lines. *British Journal of Cancer*, 111(6):1150–8.

Selinski, S., Blaszkewicz, M., Ickstadt, K., Gerullis, H., Otto, T., Roth, E., Volkert, F., Ovsiannikov, D., Moormann, O., Banfi, G., Nyirady, P., Vermeulen, S. H., Garcia-Closas, M., Figueroa, J. D., Johnson, A., Karagas, M. R., Kogevinas, M., Malats, N., Schwenn., M., Silverman, D. T., Koutros, S., Rothman, N., Kiemeney, L. A., Hengstler, J. G. and Golka, K. (2017). Identification and replication of the interplay of four genetic high-risk variants for urinary bladder cancer. *Carcinogenesis*, 38:1167–79.

Selinski, S. (2017). Discovering urinary bladder cancer risk variants: status quo after almost ten years of genome-wide association studies. *EXCLI Journal*, 16:1288–96.

Sharma, R., Yang, Y., Sharma, A., Awasthi, S. and Awasthi, Y. C. (2004). Antioxidant role of glutathione S-transferases: protection against oxidant toxicity and regulation of stress-mediated apoptosis. *Antioxidants and Redox Signaling*, 6:289–300.

Shih, A. Y., Johnson, D. A., Wong, G., Kraft, A. D., Jiang, L., Erb, H., Johnson, J. A. and Murphy, T. H. (2003). Coordinate regulation of glutathione biosynthesis and release by Nrf2-expressing glia potently protects neurons from oxidative stress. *The Journal of Neuroscience*, 23:3394–406.

Singh, M., Shah, P. P., Singh, A. P., Ruwali, M., Mathur, N., Pant, M. C. and Parmar, D. (2008). Association of genetic polymorphisms in glutathione S-transferases and susceptibility to head and neck cancer. *Mutation Research*, 638(1–2):184–94.

Singh, V., Jaiswal, P. K. and Mittal, R. D. (2014). Replicative study of GWAS TP63C/T, TERTC/T, and SLC14A1C/T with susceptibility to bladder cancer in north Indians. *Urology Oncology*, 32:1209–14.

Strange, R. C., Spiteri, M. A., Ramachandran, S. and Fryer, A. A. (2001). "Glutathione-S-transferase family of enzymes." *Mutation Research*, 482(1–2):21–6.

Surh, Y. J., Kundu, J. K. and Na, H. K. (2008). Nrf2 as a master redox switch in turning on the cellular signaling involved in the induction of cytoprotective genes by some chemopreventive phytochemicals. *Planta Medica*, 74:1526–39.

Tew, K. D., Manevich, Y., Grek, C., Xiong, Y., Uys, J. and Townsend, D. M. (2011). The role of glutathione S-transferase P in signaling pathways and S-glutathionylation in Cancer. *Free Radical Biology and Medecine*, 51(2):299–313.

Torre, L. A., Bray, F., Siegel, R. L., Ferlay, J., Lortet-Tieulent, J. and Jemal, A. (2015). Global cancer statistics, 2012. *CA: A Cancer Journal for Clinicians*, 65:87-108.

Torres-Rivera, A. and Landa, A. (2008). Glutathione transferases from parasites: a biochemical view. *Acta Tropica*, 105:99–112.

Valko, M., Leibfritz, D., Moncol, J., Cronin, M. T., Mazur, M. and Telser, J. (2007). Free radicals and antioxidants in normal physiological functions and human disease. *The International Journal of Biochemistry and Cell Biology*, 39:44–84.

Verma, A., Kapoor, R. and Mittal, R. D. (2017). Cluster of differentiation 44 (CD44) gene variants: a putative Cancer stem cell marker in risk prediction of bladder Cancer in north Indian population. *Indian Journal of Clinical Biochemistry*, 32:74–83.

Vontas, J.G., Small, G. J. and Hemingway, J. (2001). Glutathione-S-transferases as antioxidant defence agents confer pyrethroid resistance in *Nilaparvata lugens*. *Biochemical Journal*, 357:65–72.

Vontas, J. G., Small, G. J., Nikou, D. C., Ranson, H. and Hemingway, J. (20020. Purification, molecular cloning and heterologous expression of a glutathione-S-transferase involved in insecticide resistance from the rice brown planthopper, *Nilaparvata lugens*. *Biochemical Journal*, 362:329–37.

Wang, Z., Xue, L., Chong, T., Li, H., Chen, H. and Wang, Z. (2013). Quantitative assessment of the association between glutathione S-transferase P1 Ile105Val polymorphism and bladder cancer risk. *Tumor Biology*, 34:1651–7.

Wieczorek, E., Wasowicz, W., Gromadzinska, J. and Reszka, E. (2014). Functional polymorphisms in the matrix metalloproteinase genes and their association with bladder cancer risk and recurrence: a mini-review. *International Journal of Urology*, 21:744–52.

Wu, K., Wang, X., Xie, Z., Liu, Z. and Lu, Y. (2013). Glutathione S-transferase P1 gene polymorphism and bladder cancer susceptibility: an updated analysis. *Molecular Biology Reports*, 40:687–95.

Wu, G., Fang, Y.-Z., Yang, S., Lupton, J. R. and Turner, N. D. (2004). Glutathione metabolism and its implications for health. *The Journal of Nutrition*, 134:489–92.

Yang, Y., Cheng, J. Z., Singhal, S. S., Saini, M., Pandya, U., Awasthi, S. and Awasthi, Y. C. (2001). Role of glutathione S-transferases in protection against lipid peroxidation. Overexpression of hGSTA2–2 in K562 cells protects against hydrogen peroxide-induced apoptosis and inhibits JNK and caspase 3 activation. *Journal of Biological Chemistry*, 276: 19220–30.

Yang, Y., Sharma, R., Cheng, J. Z., Saini, M. K., Ansari, N. H., Andley, U. P., Awasthi, S. and Awasthi, Y. C. (2002a). Transfection of HLE B-3 cells with hGSTA1 or hGSTA2 protects against hydrogen peroxide and naphthalene induced lipid peroxidation and apoptosis. *Investigative Ophthalmology and Visual Science*, 43:434–45.

Yang, Y., Sharma, R., Zimniak, P. and Awasthi, Y. C. (2002b). Role of α class glutathione S-transferases as antioxidant enzymes in rodent tissues. *Toxicology and Applied Pharmacology*, 182:105–15.

Yin, X. and Chen, J. (2017). Is there any association between glutathione S-transferases M1 and glutathione S-transferases T1 Gene polymorphisms and endometrial Cancer risk? A Meta-analysis. *International Journal of Preventive Medicine*, 8:47.

Youssef, R. F. and Lotan, Y. (2011). Predictors of outcome of non-muscle-invasive and muscle-invasive bladder cancer. *The Scientific World Journal*, 11:369–81.

Yu, C., Hequn, C., Longfei, L., Long, W., Zhi, C., Feng, Z., Jinbo, C., Chao, L. and Xiongbing, Z. (2017). GSTM1 and GSTT1 polymorphisms are associated with increased bladder cancer risk: Evidence from updated meta-analysis. *Oncotarget*, 8(2):3246-58.

Yu, Y., Li, X., Liang, C., Tang, J., Qin, Z., Wang, C., Xu, W., Hua, Y., Shao, P. and Xu, T. (2016). The relationship between GSTA1, GSTM1, GSTP1, and GSTT1 genetic polymorphisms and bladder cancer susceptibility: A meta-analysis. *Medicine*, 95(37):e4900.

Zhang, R., Xu, G., Chen, W. and Zhang, W. (2011). Genetic polymorphisms of glutathione S-transferase M1 and bladder cancer risk: a meta-analysis of 26 studies. *Molecular Biology Reports*, 38:2491–7.

Zhang, Y., Yuan, Y., Chen, Y., Wang, Z., Li, F. and Zhao, Q. (20160. Association between GSTP1 Ile105Val polymorphism and urinary system cancer risk: evidence from 51 studies. *Onco Targets and Therapy*, 9:3565–9.

Zhang, Z., Zhang, G., Gao, Z., Li, S., Li, Z., Bi, J., Liu, X. and Kong, C. (2017). Comprehensive analysis of differentially expressed genes associated with PLK1 in bladder cancer. *BMC Cancer*, 17:861.

Zhang, Y., Munday, R., Jobson, H. E., Munday, C. M., Lister, C., Wilson, P., Fahey, J. W. and Mhawech-Fauceglia, P. (2006). Induction of GST and NQO1 in cultured bladder cells and in the urinary bladders of rats by an extract of broccoli (*Brassica oleracea* italica) sprouts. *Journal Agricultural and Food Chemistry*, 54:9370–6.

Zhou, T., Li, H. Y., Xie, W. J., Zhong, Z., Zhong, H. and Lin, Z. J. (2018). Association of Glutathione S-transferase gene polymorphism with bladder cancer susceptibility. *BMC Cancer*, 18(1):1088.

Zhu, H., Jia, Z., Misra, B., Zhang, L., Cao, Z., Yamamoto, M., Trush, M., Misra, H. and Li, Y. (2008a). Nuclear factor E2-related factor 2-dependent myocardiac cytoprotection against oxidative and electrophilic stress. *Cardiovascular Toxicology*, 8:71–85.

Zhu, H., Jia, Z., Strobl, J., Ehrich, M., Misra, H. and Li, Y. (2008b). Potent induction of total cellular and mitochondrial antioxidants and phase 2 enzymes by cruciferous sulforaphane in rat aortic smooth muscle cells: Cytoprotection against oxidative and electrophilic stress. *Cardiovascular Toxicology*, 8:115–25.

Zhu, H., Jia, Z., Zhang, L., Yamamoto, M., Misra, H. P., Trush, M. A. and Li, Y. (2008c). Antioxidants and phase 2 enzymes in macrophages: regulation by nrf2 signaling and protection against oxidative and electrophilic stress. *Experimental Biology and Medicine*, 233:463–74.

In: Glutathione S-Transferases
Editor: Igor Azevedo Silva

ISBN: 978-1-53618-188-3
© 2020 Nova Science Publishers, Inc.

Chapter 4

GLUTATHIONE S-TRANSFERASES IN INFERTILITY

Maria Manuel Casteleiro Alves[1,2], António Hélio Oliani[2], Luiza Breitenfeld[1] and Ana Cristina Ramalhinho[1,2,]*

[1]CICS-UBI – Centro de Investigação em Ciências da Saúde, Universidade da Beira Interior, Covilhã, Portugal
[2]Centro Hospitalar Cova da Beira, E.P.E., Covilhã, Portugal
Faculdade de Medicina de São José do Rio Preto (FAMERP), São José do Rio Preto, Brasil

ABSTRACT

The Glutathione S-transferases (GSTs) family plays an important role in the detoxification of environmentally toxic compounds and products of oxidative stress, neutralizing ROS production. Oxidative stress is referred as the imbalance between oxidants and antioxidants and the generation of excessive amounts of reactive oxygen species (ROS). In

---

* Corresponding Author's E-mail: cramalhinho@fcsaude.ubi.pt.

a healthy body, ROS and antioxidants remain in balance. When the balance is disrupted towards an overabundance of ROS, oxidative stress occurs. The presence of unbalanced ROS can cause cellular damage and change cellular functions because they regulate protein activity and gene expression, which can lead to several effects. Oxidative stress is recognized to play a central role in the pathophysiology of many different disorders, including infertility. Infertility is a disease characterized by the failure to establish a clinical pregnancy after 12 months of regular, unprotected sexual intercourse or due to an impairment of a person's capacity to reproduce either as an individual or with his/her partner. The incidence of infertility differs between racial and ethnic groups because of the multifactorial nature of the disease. The etiology and pathogenesis of infertility are still unclear. However, there is increasing evidence that infertility depends on complex interactions between genetic factors and environmental toxins which can be implicated in its pathogenesis. Because of GSTs detoxification properties, it is logical to suspect that dysfunction of detoxification-related enzymes might be a contributor to the development of infertility. In this chapter, we will review the association of GSTs common genetic variants with the development of infertility in women and in men.

Keywords: infertility, GSTs, oxidative stress, ROS

INTRODUCTION

Infertility is among the most serious medical problems worldwide. Approximately 80 million people worldwide are affected by infertility (Showell et al., 2014; Macanovic et al., 2015; Wang et al., 2017; Dobrakowski et al., 2018). Worldwide, about 1 in 10 couples develop primary or secondary infertility. Infertility is a disease characterized by the failure to establish a clinical pregnancy after 12 months of regular, unprotected sexual intercourse or due to an impairment of a person's capacity to reproduce either as an individual or with his/her partner. Fertility interventions may be initiated in less than one year based on medical, sexual and reproductive history, age, physical findings and diagnostic testing (Zegers-Hochschild et al., 2017; Borght et al., 2018). The etiology and pathogenesis of infertility are still unclear. However, there is increasing evidence that infertility depends on complex interactions between genetic factors and environmental which can be implicated in its pathogenesis (Kennedy et al., 2003; Rier et al., 2008; Henidi et al., 2014).

The incidence of infertility differs between ethnic groups because of the multifactorial nature of the disease. Infertility is a couple disease which may have female, male or mixed causes. Several disorders are known causes for male infertility such as oligozoospermia; asthenozoospermia; azoospermia; teratozoospermia; and varicocele. Oligozoospermia refers to the reduction in the amount of sperm present in the ejaculated fluid. Asthenozoospermia implies the reduction of sperm motility and azoospermia is characterized by the absence of any sperm in the ejaculated fluid. The term teratozoospermia is used when there is a large percentage of abnormally shaped sperm (Xiong et al., 2015). Varicocele is the abnormal tortuosity and dilatation of the veins of the pampiniform plexus within the spermatic cord. The exact pathophysiology of varicocele remains unknown (Goldstein et al., 1989; Gorelick et al., 1993; Shiou-Sheng et al., 2002; Zhu et al., 2014). However, several studies have also suggested that this condition is associated with increased oxidative stress (Saleh et al., 2003; Allamaneni et al., 2004; Mancini et al., 2004; Xiong et al., 2015).

Female infertility has also several infertility factors; we can highlight endometriosis, polycystic ovary syndrome (PCOS) and premature ovarian failure (POF). Also, tubal factors and ovulatory events and disorders can contribute to infertility.

Endometriosis is defined as the presence of endometrial tissue outside the uterine cavity. Increasing evidence suggests that endometriosis is a polygenic and multifactorial disease (Eskenazi et al., 1997; Henidi et al., 2014; Batista et al., 2017). Several candidate genes have been investigated to specify the women at increased risk of endometriosis or to understand its pathogenesis (Thomas et al., 2000; Henidi et al., 2014; Batista et al., 2017). PCOS is a hormonal disorder, affecting women at reproductive age and associated with reproductive, metabolic, and psychological dysfunction. Women with this disease commonly have features of hyperandrogenism and hirsutism, oligo or amenorrhea, and anovulation. This disorder occurs in 5-10% women in reproductive ages (Mohammadi et al., 2019). Despite a long history of studies on PCOS, its etiology is still unknown (Moran et al., 2010; Zhang et al., 2012; Murri et al., 2013; Mohammadi et al., 2019). POF is a condition when ovaries stop working long before the expected menopausal time. Some of the diagnostic symptoms of the disease include amenorrhea and hypoestrogenism. The cause of POF in most cases is idiopathic (Ghahremani-Nasab et al., 2019). Up to a third of couples are diagnosed with *unexplained infertility,* or sterility of unknown cause. In *unexplained infertility* abnormalities are likely to be present but cannot be detected by current methods (Safarinejad et al., 2012).

GLUTATHIONE S-TRANSFERASES (GSTs)

Glutathione S-transferases (GSTs) are a family of multifunctional enzymes that catalyze the conjugation of oxidative stress products, environmental toxins, carcinogens and reactive electrophiles, inactivating them by binding to glutathione. The electrophilic compounds become more soluble when conjugated to glutathione, so can be easier and quickly excreted (Fafula et al., 2019).

Human GSTs are divided into three distinct super family members: mitochondrial, cytosolic and microsomal, currently named Membrane Associated Proteins in Eicosanoid and Glutathione Metabolism (MAPEG) GSTs. Members of the GST superfamily are extremely diverse in amino acid sequence, and a large fraction of the sequences deposited in public databases are of unknown function (Di Pietro et al., 2010). Human cytosolic GSTs represent the largest and most complex family and they are considered the most relevant to disease research (Hayes et al., 2005). Human cytosolic GSTs are divided into seven classes: Alpha, Mu, Omega, Pi, Sigma Theta and Zeta, respectively encoded by genes GSTA, GSTM, GSTO, GSTP, GSTS, GSTT and GSTZ (Safarinejad et al., 2012; Nair et al., 2013). These GSTs display polymorphisms that may contribute to interindividual differences in responses to xenobiotics and reactive electrophiles. Oxygen toxicity is an inherent challenge to aerobic life. ROS can modulate cellular functions, and oxidative stress can impair the intracellular milieu resulting in diseased cells or endangered cell survival. In the healthy human body, ROS and antioxidants remain in balance. When the balance is interrupted, increasing the amount of ROS, oxidative stress occurs (Agarwal et al., 2005). Thus, GSTs genotypes, alone or in combination, may determine subjects as "low metabolizers", and consequently more likely to suffer formation of DNA mutations that confer major susceptibility to complex multifactorial diseases, with genetic and environmental influences (Di Pietro et al., 2010).

To understand the relationship between GSTs and disease, several studies have been conducted. In infertility, the most studied genes are GSTM, GSTT and GSTP genes. In humans, deletion polymorphisms of the genes encoding GSTM1 and GSTT1 are common (Josephy, 2010). The GSTM1 gene locates on chromosome 1p13.3 and has a common functional variant (null versus present). The frequency of this deletion is between 23% and 63% depending on the population studied (Safarinejad et al., 2012). The GSTM1 are widely expressed in all human tissues including in ovary, fallopian tube, endometrium, uterus, testis, epididymis, seminal vesicle and prostate gland (ENSG00000134184). The GSTT is divided into two subunits, GSTT1 and GSTT2 and are both located on chromosome 22q11. GSTT1 are mainly expressed in lung and liver, but are also expressed in prostate gland and placenta (ENSG00000277656). GSTT2 are mainly expressed in stomach, adrenal gland, esophagus and pancreas, but are also expressed in ovary, uterus, fallopian tube and prostate gland (ENSG00000099984). In both cases, a gene deletion is responsible for the existence of a null allele (Kan et al., 2013). These deletions presumably arose by homologous recombination events. The designated GSTM1 and GSTT1 null genotypes have demonstrated functional relevance, that is, inexistence of enzymatic activity (Kan et al., 2013). Individuals with a homozygous deletion in GSTM1 and GSTT1 locus show no functional enzymatic activity of cytosolic enzyme (Safarinejad et al., 2012). The GSTP class is encoded by a single gene, GSTP1, located on chromosome 11q13. GSTP1 are mainly expressed in thyroid gland, lung, liver, epididymis, seminal vesicle and fallopian tube, but is also expressed in ovary, endometrium, uterus, testis and prostate gland (ENSG00000084207). GSTP1 is polymorphic with two common functional variants based on substitutions in amino acids 105, Isoleucine (Ile) to Valine (Val), and 114, Alanine (Ala) to Val. Thus, four haplotypes have been identified: the wild-type GSTP1*A (Ile105 + Ala114) and three variant haplotypes, GSTP1*B (Val105 + Ala114), GSTP1*C (Val105 + Val114) and GSTP1*D (Ile105 + Val114). The Ala114Val polymorphism seems not to influence the enzyme activity (Ali-Osman et al., 1997; Ramalhinho et al., 2011).

GST AND MALE INFERTILITY

Male infertility is a worldwide polygenic complicated disease (Boivin et al., 2007; Safarinejad et al., 2008; Xiong et al., 2015; Krausz et al., 2018) and represents a typical example of complex multifactorial disease with a substantial genetic basis (Ferlin, 2012; Xiong et al., 2015). Genetic deficiency, infection, varicocele, immunological problems, and environmental factors could affect sperm quality or function (Krausz et al., 2018). Reactive oxygen species (ROS) from pollution, radiation, high-fat diets and sedentary physically inactive lifestyles will likely contribute to the increase in incidence of male infertility (Storgaard et al., 2006; Yu et al., 2015). Xenobiotics and endogenous ROS are the major factors leading to oxidative stress (OS) in the male testis tissue. So, high concentrations of ROS can cause potential damage to plasma membrane and DNA integrity, motility, viability, fertilization ability and overall semen quality (Aitken et al., 2002; Fafula et al., 2019). Male germ cells are particularly sensitive to OS, which could contribute to testicular and sperm dysfunctions as well as toxicity to spermatogenesis (Aitken et al., 2001; Agarwal et al., 2005; Tremellen et al., 2008; Turner et al., 2008). OS could lead to such biological effects as the acceleration of spermatozoa apoptosis (Aitken et al., 2012), abnormality of sperm quality parameters (Badade et al., 2011), decrease of sperm and oocyte fusion capacity (Griveau et al., 1997; Aitken et al., 1998; Aitken et al., 2012), and damage of DNA integrity in both sperm mitochondrial and nuclear genomes (Aitken et al., 1998; Aitken et al., 2012). Oxidative damage to the sperm membranes leads to reduced ejaculate quality and has been considered as one of the causes of male infertility (Tremellen et al., 2008; Tafuri et al., 2015; Fafula et al., 2019). As was said earlier, OS is an imbalance between ROS production and the semen's natural antioxidant defenses (Showell et al., 2014); therefore to ensure fertility, the process to control oxidative stress seems to be important.

The antioxidants play an important protective role on male germ cells against OS (Schuppe et al., 2000; Agarwal et al., 2006; Showell et al., 2014). GSTs constitute the major defensive antioxidant system against oxidative stress because are involved in the detoxification of both endogenous and exogenous substances (Dusinská et al., 2001; Xiong et al., 2015). They are present in the testis and seminiferous tubule fluid as well as in the sperm (Mukherjee et al., 1999; Hemachand et al., 2002). GSTA, GSTM, GSTT, and GSTP act as important antioxidants in testis tissues (Listowsky et al., 1998; Strange et al., 2001; Fafula et al., 2019), protecting Leydig cells, Sertoli cells, and germ cells against the damage induced by OS (Castellon, 1999; Hemachand et al., 2002). GSTs are also present in female reproductive tract, and sperm cells might use this fluid as a source for GSTs to protect them from OS (Xiong et al., 2015).

A meta-analysis (Wu W. et al., 2013) including 3981 cases and 2953 controls from 19 case-control studies, showed that GSTM1 null genotype was significantly associated with male infertility risk, which was consistent with four previous meta-analysis (Li et al., 2012; Safarinejad et al., 2012; Tang M. et al., 2012; Xu et al., 2013). In the same study, for GSTT1 null genotype, a significant association with male infertility risk was only found among Asians (Wu W. et al., 2013). Another meta-analysis supported a significant association between GSTM1 null genotype and risk of male infertility especially in Caucasians (Chengyong et al., 2012). The meta-analysis of Ying et al., demonstrated that the null genotype of GSTM1 might be a potential factor provoking male infertility in Caucasian and Chinese populations, while null genotype of GSTT1 was related to male infertility only in Chinese population (Ying et al., 2013). Also, an increase in infertility risk was described in Iranian (Finotti et al., 2009) and Brazilian (Safarinejad et al., 2010) populations.

Several studies demonstrated an association between male infertility and the GSTM1 and GSTT1 null deletion in men with idiopathic infertility (Aydemir et al., 2007; Tirumala et al., 2010; Polonikov et al, 2010; Salehi et al., 2012; Xu et al., 2013). A study described specifically an association between the GSTT1 null genotype and a reduction in sperm concentration and count (Olshan et al., 2010). These results suggest that GST gene polymorphisms has a relationship with male infertility, but the exact molecular mechanisms of GSTM1 and GSTT1 null polymorphisms on male infertility still unclear.

Regarding GSTP1, a report found that the variant genotypes of GSTP1 (Ile/Val + Val/Val) resulted in a significant decreased risk of infertility (Safarinejad et al., 2010). On the other hand, Tang et al. found that GSTP1 allelic variation showed barely any difference between the infertile and control groups (Tang K. et al., 2012). Another report, revealed that all individuals in both the oligoasthenoteratozoospermia and normozoospermia groups had the same GSTP1 (Ile/Ile) genotype, indicating no significant association between GSTP1 and sperm parameters (Lakpour et al., 2013). While several studies suggested that the GSTM1 and GSTT1 null genotypes are associated with an increase risk of male infertility, as seen before, the GSTP1 Ile/Val genotype seems to have a protective effect (Safarinejad et al., 2012). However, a study in China, that did not found significant association between GSTM1 and GSTT1 null polymorphisms with the risk of develop male infertility, suggested that the GSTP1 variant genotype (Ile/Val + Val/Val) may be associated to the susceptibility to develop infertility. In this study was also reported that the combined effect of the GSTT1 null genotype and GSTP1 (Ile/Val + Val/Val) is associated with an increased risk of develop infertility. In this same study the GSTM1 null deletion had barely any effect when the three GST genes were studied together (Xiong et al., 2015).

GSTT1 null genotype seems to predispose the spermatocytes of infertile males with varicocele to excess oxidative damage (Wu et al., 2008). The sperm of varicocele patients with GSTM1 null genotypes also appear to be more vulnerable to oxidative damage (Shiou-Sheng et al., 2002), so should be paid more attention to oxidative stress related pathological manifestations in this patients. A meta-analysis, including 497 cases and 476 controls in an overall population analysis reported that GSTM1 and GSTT1 null polymorphisms were not observed more frequently in varicocele patients than in controls. Only in the Caucasian subgroup analysis was found a significantly higher frequency of the GSTM1 null genotype amongst varicocele patients. Such association was not reported in the Asian subgroup. This meta-analysis also did not find any statistically significant association concerning the GSTT1 null genotype in either Caucasian or Asian populations. Contrasting with these findings, a study in Turkey revealed no statistically significant association between GSTM1 and GSTT1 null deletion and development of varicocele (Acar et al., 2011). In other perspective, a study reported that the response rate to varicocelectomy is significantly higher in patients with the GSTT1 presence genotype than the GSTT1 null genotype and the response rate became higher in combination with the GSTM1 present genotype (Okubo et al., 2005). A study from Egypt described that internal spermatic vein of infertile men associated with varicocele has decreased levels of GST compared with peripheral venous circulation, what is correlated with both oxidative stress and varicocele grade (Mostafa et al., 2015). So, the impact of varicocele on male fertility remains unknown (Baazeem et al., 2011) and because of this, future studies needs should include other genetic polymorphisms of varicocele patients.

GST AND FEMALE INFERTILITY

Oxidative stress influences the entire reproductive lifespan of a woman. ROS have a physiological and pathological role in the female reproductive tract. Several studies have reported the presence of ROS in the female reproductive tract: ovaries (Shiotani et al., 1991; Jozwik et al., 1999; Sabatini et al., 1999; Behrman et al., 2001); fallopian tubes (El Mouatassim et al., 1999) and embryos (Guérin et al., 2001). ROS are involved in the modulation of an entire spectrum of physiological reproductive functions such as oocyte maturation, ovarian steroidogenesis, corpus luteal function and luteolysis (Sabatini et al., 1999; Behrman et al., 2001).

Several studies have investigated the role of ROS in various diseases of the female reproductive tract, concluding that ROS can affect a variety of physiological functions in the reproductive tract, and excessive levels can result in precipitous pathologies affecting female reproduction (Agarwal et al., 2005).

The acknowledged importance of GSTs in the disposition of toxic compounds and in defense against oxidative stress (Agarwal et al., 2005) justifies testing associations with the risk of various diseases, including infertility. The absence of activity of GSTM1 and GSTT1 proteins due to deletion of GSTM1 and GSTT1 genes may facilitate and increase the production of ROS, promoting DNA damage and apoptosis, and impeding the proliferation and invasion of ectopic endometrial cells (Ehrmann, 2005; Kubiszeski et al., 2015).

Several studies in various populations have shown a highly significant association of GSTM1 and GSTT1 null genotypes with endometriosis (Baranova et al., 1999; Lin et al., 2003; Hosseinzadeh et al., 2011; Ding et al., 2014; Henidi et al., 2014; Zhu et al., 2014; Chen et al., 2015; Li et al., 2015; Hassani et al., 2016; Xin et al., 2016). In contrast, few studies did not observe an association of GSTM1 and GSTT1 null genotype with endometriosis (Baxter et al., 2001; Seifati et al., 2012; Matsuzaka et al., 2012; Vichi et al., 2012; Kubiszeski et al., 2015; Batista et al., 2017). Regarding GSTP1 polymorphisms, two reports described that they do not affect the pathogenesis of endometriosis (Hur et al., 2005; Tuo et al., 2016). However, one study mentioned that GSTP1 polymorphisms might modulate the risk of endometriosis with significantly decreased risk for GSTP1 Val/Val and a trend for increased risk for GSTP1 Ile/Ile (Ertunc et al., 2005).

Several pathways have been implicated in the etiology of PCOS, some of which included the metabolic or regulatory pathways of steroid hormone synthesis (Escobar-Morreale et al., 2005; Fratantonio et al., 2005; Henidi et al., 2014). Elevated steroid hormone concentrations are seen in women with polycystic ovaries (Greisen et al., 2001). GSTs play an important role in steroid hormone metabolism. In addition, they are also implied in the maintenance of cellular redox potential and imposing oxidative stress. These factors play a crucial role in the programmed development of a single follicle inside the ovary, whose disturbance can lead to the formation of cysts in the ovary (Babu et al., 2004). The only study, to our knowledge, that evaluated the association of GSTM1 null genotype with PCOS, did not corroborate the hypothesis that GSTM1 null genotype is significantly associated with this disease (Babu et al., 2004).

To our knowledge, there are no reports correlating GSTs with POF. However, was studied that high superoxide ion levels lead to a decrease in the bioavailability of nitric oxide and an increase in ROS levels and oxidative stress (Lu et al., 2008). As compared to spermatozoa, female germ cells develop under hypoxic condition in the ovarian cortex, however exposure to supraphysiological levels of ROS are detrimental to developing oogonia (Coulam et al., 1986). It was suggested that increased production of ROS contributes to oophoritis associated with premature ovarian insufficiency (Behrman et al., 2001). GSTM1 and GSTT1 genes deletions, that suppress the coding of GSTM1 and GSTT1 proteins, may have a role in the maintenance of high ROS levels, as the conjugation of oxidative stress products and reactive electrophiles may be compromised. High ROS levels induce mitochondrial DNA alterations and lead to mitochondria dysfunction (Kumar et al., 2010) what could lead to low production of adenosine triphosphate (ATP) due to impaired oxidative phosphorylation and thus to impaired oogenesis, low oocyte number and POF (Manoj et al., 2012). Some reports have already stated that oxidative stress is implicated in the pathophysiology of tubal factor infertility (Agarwal et al., 2005; Polak et al., 2011). Postulated pathologic mechanisms include oxidative DNA damage, lipid peroxidation, modulation of gene expression and transcription factors, inhibition of protein synthesis, and depletion of ATP (Ray et al., 2004).

THE ROLE OF GST ON EMBRYONIC DEVELOPMENT

The oxidant status can influence early embryo development by modifying the key transcription factors and hence modifying gene expression (Whitbread et al., 2003). Concentrations of ROS may also play a major role both in oocyte fertilization and embryo implantation (Zhao et al., 2009). As oocyte quality and its microenvironment affect early embryo development (Krisher et al., 2004), many studies have tried to identify biomarkers for the oocyte microenvironment, to be used as predictive factors of embryo and pregnancy outcomes (Lédée et al., 2013; Uyar et al., 2013).

Several indicators of oocyte quality have been reported using cumulus-oocyte complexes (COCs) from infertile patients, taking into account that these are important in achieving a successful pregnancy. GSTs are present in the oocyte and tubal fluid and have a role in improving the development of the zygote beyond the 2-cell block to the morula or the blastocyst stage (Agarwal et al., 2005). In this context, GSTT1 was studied as a possible predictor of oocyte quality. However, one study revealed that GSTT1 expression in the samples from tubal and unknown factor patients did not correlate with COC maturity. Furthermore, the developmental capacity of oocytes from the selected patients with low GSTT1 was likely to be higher than that with high GSTT1 (Ito et al., 2008). On the other hand, a report described that the accumulation of products of oxidative stress inducing digestion of DNA, in mural and cumulus granulosa cells from infertile patients are correlated negatively with the oocyte quality (Seino et al., 2002). In agreement with this finding, a study mentioned that the amount of ROS in human follicular fluid was negatively correlated with oocyte development potential (Tsai-Turton et al., 2006). Moreover, another study related that apoptosis possibly induced by ROS in granulosa cells from infertile patients was closely related to the oocyte quality (Nakahara et al., 1997). A report from Spain revealed that GST activity was significantly lower in mature oocytes compared to immature. These results suggest that GST may play a role in the follicle maturation by detoxifying xenobiotics, thus contributing to the normal development of oocyte (Meijide et al., 2014; Olszak-Wąsik et al., 2019)

Some authors reported that ovarian stimulation efficiency in infertile patients and pregnancy success is correlated with elevated total antioxidant concentration (Velthut et al., 2014). In agreement with this study, another report described that the glutathione system is strongly engaged in maintaining intra and extracellular redox balance (Erel, 2004; Agarwal et al., 2005; Dalto et al., 2017). Another report has shown that the follicular fluid concentrations of GST enzymes are directly associated with higher oxidative stress levels in young patients undergoing ovarian stimulation cycles compared with fertile oocyte donors and highresponse patients (Nuñez-Calonge et al., 2016). A study in Poland indicated that carriers with higher GSTs have a decreased embryo quality and is associated with lower pregnancy chance (Olszak-Wasik et al., 2019). A report suggested that GST is involved in the normal development of oocyte and follicle maturation with lower activity found in mature oocytes (compared to immature ones), both in donors and IVF patients (Meijide et al., 2014; Olszak-Wasik et al., 2019).

So, it is necessary to highlight that early preimplantation development of the embryo is strongly dependent on oocyte quality, especially up to the 3rd day of development (Tesarik et al., 2004). Thus, the pool of antioxidants stored in oocyte during oogenesis creates the system used by embryo to defend against ROS (El Mouatassim et al., 1999).

CONCLUSION

In the present chapter, we provide an overview of GSTs in infertility. The importance of glutathione is evident by its extensive expression in human body. GSTs are antioxidant enzymes that play an important role in the detoxification of oxidative stress products, environmental toxins, carcinogens and reactive electrophiles, inactivating them by binding to glutathione. Allelic variants of relevant xenobiotic metabolizing enzymes are often considered as a differential risk of developing various diseases such as infertility. Most studies emphases polymorphisms in GSTM1 and GSTT1 because they may suffer homozygous deletions. Thus, populations with homozygous deletions of GSTM1 or GSTT1 may be at greater risk for developing diseases due to their impaired ability to reduced detoxification efficiency. Several epidemiological studies confirmed that GSTM1 and GSTT1 deletions are correlated with an increased susceptibility to diseases associated with OS (Bolt et al., 2006; Bohanec et al., 2009). The possible role in male and female infertility has been suggested for GSTM1 and GSTT1 gene variants, whereas published data are inconsistent. The inconsistent results between studies reflect the complexity in the role of GSTs and might be due, for example, because some GSTs isoforms like GSTP1 may exhibit different activity, affinity, and thermostability according to genotype and substrates. Furthermore, GSTs are known to have overlapping substrate specificities and the absence of GST isoenzymes may be compensated by other isoforms. In fact, a study in India revealed that the absence of GSTM1 activity can be compensated by the overexpression of GSTM2 (Bhattacharjee et al., 2013), so one may infer that other members of the GST family might compensate for the loss of GSTM1, GSTT1 and GSTP1 activity. Also, different populations with different risk agent exposures may explain the differences in the outcomes of the reported studies conducted on this topic. Consequently, further studies with larger samples sizes, and considering gene–environmental and gene–gene interactions analysis should be performed.

REFERENCES

Acar, H; Kilinç, M; Guven, S; Inan, Z. Glutathione S-transferase M1 and T1 polymorphisms in Turkish patients with varicocele. *Andrologia.*, 2011, 44(1), 34–37.

Agarwal, A; Gupta, S; Sharma, RK. Role of oxidative stress in female reproduction. *Reprod Biol Endocrinol.*, 2005, 14, 3, 28.

Agarwal, A; Gupta, S; Sikka, S. The role of free radicals and antioxidants in reproduction. *Curr Opin Obstet Gynecol.*, 2006, 18, 325–332.

Agarwal, A; Sushil, P. Oxidative stress and antioxidant in male infertility: a difficult balance. *Iranian J Reprod Med.*, 2005, 3, 1–8.

Aitken, RJ; De Iuliis, GN; Gibb, Z; Baker, MA. The simmet lecture: new horizons on an old landscape-oxidative stress, DNA damage and apoptosis in the male germ line. *Reprod Domest Anim.*, 2012, 47(Suppl 4), 7–14.

Aitken, RJ; Gordon, E; Harkiss, D; Twigg, JP; Milne, P; Jennings, Z; Irvine, DS. Relative impact of oxidative stress on the functional competence and genomic integrity of human spermatozoa. *Biol Reprod.*, 1998, 59(5), 1037–1046.

Aitken, RJ; Krausz, C. Oxidative stress, DNA damage and the Y chromosome. *Reproduction.*, 2001, 122, 497–506.

Aitken, RJ; Marshall Graves, JA. The future of sex. *Nature.*, 2002, 415, 963.

Ali-Osman, F; Akande, O; Antoun, G; Mao, J; Buolamwini, J. *J. Biol. Chem.*, 1997, 15, 10004

Allamaneni, SS; Naughton, CK; Sharma, RK; Thomas, AJ; Jr. Agarwal, A. Increased seminal reactive oxygen species levels in patients with varicoceles correlate with varicocele grade but not with testis size. *Fertil Steril.*, 2004, 82(6), 1684–1686.

Aydemir, B; Onaran, I; Kiziler, AR; Alici, B; Akyolcu, MC. Increased oxidative damage of sperm and seminal plasma in men with idiopathic infertility is higher in patients with glutathione S-transferase Mu-1 null genotype. *Asian journal of andrology.*, 2007, 9, 108–115.

Baazeem, A; Belzile, E; Ciampi, A; Dohle, G; Jarvi, K; Salonia, A; Weidner, W; Zini, A. Varicocele and male factor infertility treatment: a new meta-analysis and review of the role of varicocele repair. *Eur Urol.*, 2011, 60(4), 796–808.

Babu, AK; Rao, LK; Kanakavalli, MK; Suryanarayana, VV; Deenadayal, M; Singh, L. CYP1A1, GSTM1 and GSTT1 genetic polymorphism is associated with susceptibility to polycystic ovaries in South Indian women. *Reproductive BioMedicine Online.*, 2004, 9, 2, 194-200.

Badade, G; More, K; Narshetty, G; Badade, V; Yadav, B. Human seminal oxidative stress: correlation with antioxidants and sperm quality parameters. *Ann Biol Res.*, 2011, 2, 351–359.

Baranova, H; Canis, M; Ivaschenko, T; Albuisson, E; Bothorishvilli, R; Baranov, V; Malet, P; Bruhat, MA. Possible involvement of arylamine N-acetyltransferase 2, glutathione S-transferases M1 and T1 genes in the development of endometriosis. *Mol Hum Reprod.*, 1999, 5(7), 636-41.

Batista, BC; Ruiz-Cintra, MT; Lima, MC; Marqui, ABT. No association between glutathione S-transferase M1 and T1 gene polymorphisms and susceptibility to endometriosis. *J Bras Patol Med Lab.*, 2017, 53, 3, 183-187.

Baxter, SW; Thomas, EJ; Campbell, IG. GSTM1 null polymorphism and susceptibility to endometriosis and ovarian cancer. *Carcinogenesis.*, 2001, 22, 63–5.

Behrman, HR; Kodaman, PH; Preston, SL; Gao, S. Oxidative stress and the ovary. *J Soc Gynecol Investig.*, 2001, 8(Suppl), S40–2.

Bhattacharjee, P; Paul, S; Banerjee, M; Patra, D; Banerjee, P; Ghoshal, N; Bandyopadhyay, A; Giri, AK. Functional compensation of glutathione S-transferase M1 (GSTM1) null by another GST superfamily member, GSTM2. *Sci Rep.*, 2013, 3, 2704.

Bohanec Grabar, P; Logar, D; Tomsic, M; Rozman, B; Dolzan, V. Genetic polymorphisms of glutathione S-transferases and disease activity of rheumatoid arthritis. *Clin Exp Rheumatol.*, 2009, 27, 229–236.

Boivin, J; Bunting, L; Collins, JA; Nygren, KG. International estimates of infertility prevalence and treatment-seeking: potential need and demand for infertility medical care. *Hum Reprod.*, 2007, 22, 1506–12.

Bolt. HM; Their. R. Relevance of the deletion polymorphisms of the glutathione S-transferases GSTT1 and GSTM1 in pharmacology and toxicology. *Curr Drug Metab.*, 2006, 7, 613–628.

Castellon, EA. Influence of age, hormones and germ cells on glutathione S-transferase activity in cultured Sertoli cells. *Int J Androl.*, 1999, 22, 49–55.

Chen, XP; Xu, DF; Xu, WH; Yao, J; Fu, SM. Glutathione-S-transferases M1/T1 gene polymorphisms and endometriosis: a meta-analysis in Chinese populations. *Gynecol Endocrinol.*, 2015, 31(11), 840–845.

Chengyong, W; Man, Y; Mei, L; Liping, L; Xuezhen, W. GSTM1 null genotype contributes to increased risk of male infertility: a meta-analysis. *J Assist Reprod Genet.*, 2012, 29(8), 837-45.

Coulam, CB; Adamson, SC; Annegers, JF. Incidence of premature ovarian failure. *Obstet Gynecol.*, 1986, 67, 604–6.

Dalto, DB; Matte, JJ. Pyridoxine (Vitamin B_6) and the Glutathione Peroxidase System; a Link between One-Carbon Metabolism and Antioxidation. *Nutrients.*, 2017, 24, 9(3), 189.

Di Pietro, G; Magno, LA; Rios-Santos, F. Glutathione S-transferases: an overview in cancer research. *Expert Opin Drug Metab Toxicol.*, 2010, 6(2), 153–170.

Ding, B; Sun, W; Han, S; Cai, Y; Ren, M. Polymorphisms of glutathione S-transferase M1 (GSTM1) and T1 (GSTT1) and endometriosis risk: a meta-analysis. *Eur J Obstet Gynecol Reprod Biol.*, 2014, 183, 114-20.

Dobrakowski, M; Kaletka, Z; Machoń-Grecka, A; Kasperczyk, S; Horak, S; Birkner, E; Zalejska-Fiolka, J; Kasperczyk, A. The Role of Oxidative Stress, Selected Metals, and Parameters of the Immune System in Male Fertility. *Oxid Med Cell Longev.*, 2018, 6249536.

Dusinská, M; Ficek, A; Horská, A; Raslová, K; Petrovská, H; Vallová, B; Drlicková, M; Wood, SG; Stupáková, A; Gasparovic, J; Bobek, P; Nagyová, A; Kovácikova, Z; Blazícek, P; Liegebel, U; Collins, AR. Glutathione S-transferase polymorphisms influence the level of oxidative DNA damage and antioxidant protection in humans. *Mutat Res.*, 2001, 482, 47–55.

Ehrmann, DA. Polycystic ovary syndrome. *N Engl J Med.*, 2005, 352, 1223–1236.

El Mouatassim, S; Guérin, P; Ménézo, Y. Expression of genes encoding antioxidant enzymes in human and mouse oocytes during the final stages of maturation. *Mol Hum Reprod.*, 1999, 5(8), 720–725.

Erel, O. A novel automated method to measure total antioxidant response against potent free radical reactions. *Clin Biochem.*, 2004, 37(2), 112–119.

Ertunc, D; Aban, M; Tok, EC; Tamer, L; Arslan, M; Dilek, S. Glutathione-S-transferase P1 gene polymorphism and susceptibility to endometriosis. *Hum Reprod.*, 2005, 20, 2157–2161.

Escobar-Morreale, HF; Luque-Ramirez, M; San Millan, JL. The molecular-genetic basis of functional hyperandrogenism and the polycystic ovary syndrome. *Endocr Rev.*, 2005, 26, 251–282.

Eskenazi, B; Warner, ML. Epidemiology of endometriosis. *Obstet Gynecol Clin North Am.*, 1997, 24(2), 235–258.

Fafula, RV; Paranyak, NM; Besedina, AS; Vorobets, DZ; Iefremova, UP; Onufrovych, OK; Vorobets, ZD. Biological Significance of Glutathione S-Transferases in Human Sperm Cells. *J Hum Reprod Sci.*, 2019, 12(1), 24–28.

Ferlin, A. New genetic markers for male fertility. *Asian J Androl.*, 2012, 14, 807–8.

Finotti, AC; Costa, E; Silva, RC; Bordin, BM; Silva, CT; Moura, KK. Glutathione S-transferase M1 and T1 polymorphism in men with idiopathic infertility. *Genet Mol Res.*, 2009, 8, 1093–8.

Fratantonio, E; Vicari, E; Pafumi, C; Calogero, AE. Genetics of polycystic ovarian syndrome. *Reprod Biomed Online.*, 2005, 10, 713–720.

Ghahremani-Nasab, M; Ghanbari, E; Jahanbani, Y; Mehdizadeh, A; Yousefi, M. Premature ovarian failure and tissue engineering. *J Cell Physiol.*, 2019, 10.1002/jcp.29376.

Goldstein, M; Eid, JF. Elevation of intratesticular and scrotal skin surface temperature in men with varicocele. *J Urol.*, 1989, 142(3), 743–745.

Gorelick, JI; Goldstein, M. Loss of fertility in men with varicocele. *Fertil Steril.*, 1993, 59(3), 613–616.

Greisen, S; Ledet, T; Ovesen, P. Effects of androstenedione, insulin and luteinizing hormone on steroidogenesis in human granulosa luteal cells. *Human Reproduction.*, 2001, 16, 2061–2065.

Griveau, GF. Le Lannou D. Reactive oxygen species and human spermatozoa: physiology and pathology. *Int J Androl.*, 1997, 20, 61–69.

Guérin, P; El Mouatassim, S; Ménézo, Y. Oxidative stress and protection against reactive oxygen species in the pre-implantation embryo and its surroundings. *Hum Reprod Update.*, 2001, 7(2), 175–189.

Hassani, M; Saliminejad, K; Heidarizadeh, M; Kamali, K; Memariani, T; Khorram Khorshid, HR. Association study of glutathione S-transferase polymorphisms and risk of endometriosis in an Iranian population. *Int J Reprod Biomed* (Yazd)., 2016, 14(4), 241-6.

Hayes, JD; Flanagan, JU; Jowsey, IR. Glutathione transferases. *Annu Rev Pharmacol Toxicol.*, 2005, 45, 51–88.

Hemachand, T; Gopalakrishnan, B; Salunke, DM; Totey, SM; Shaha, C. Sperm plasma membrane associated glutathione S-transferases as gamete recognition molecules. *J Cell Sci.*, 2002, 115, 2053–2065.

Henidi, B; Kaabachi, S; Mbarik, M; Zhioua, A; Hamzaoui, K. Glutathione S- transferase M1 and T1 gene polymorphisms and risk of endometriosis in Tunisian population. *Human Fertility.*, 2014, 18(2), 128-33.

Hosseinzadeh, Z; Mashayekhi, F; Sorouri, ZZ. Association between GSTM1 gene polymorphism in Iranian patients with endometriosis. *Gynecol Endocrinol.*, 2011, 27(3), 185-9.

Hur, SE; Lee, JY; Moon, HS; Chung, HW. Polymorphisms of the genes encoding the GSTM1, GSTT1 and GSTP1 in Korean women: No association with endometriosis. *Mol Hum Reprod.*, 2005, 11, 15–9.

Ito, M; Muraki, M; Takahashi, Y; Imai, M; Tsukui, T; Yamakawa, N; Nakagawa, K; Ohgi, S; Horikawa, T; Iwasaki, W; Iida, A; Nishi, Y; Yanase, T; Nawata, H; Miyado, K; Kono, T; Hosoi, Y; Saito, H. Glutathione S-transferase theta 1 expressed in granulosa cells as a biomarker for oocyte quality in age-related infertility. *Fertil Steril*, 2008, 90(4), 1026–1035.

Josephy, PD. Genetic variations in human glutathione transferase enzymes: significance for pharmacology and toxicology. *Hum Genomics Proteomics.*, 2010, 876–940.

Jozwik, M; Wolczynski, S; Szamatowicz, M. Oxidative stress markers in preovulatory follicular fluid in humans. *Molecular Human Reproduction.*, 1999, 5, 409–413.

Kan, HP; Wu, FL; Guo, WB; Wang, YZ; Li, JP; Huang, YQ; Li, SG; Liu, JP. Null genotypes of GSTM1 and GSTT1 contribute to male factor infertility risk: a meta-analysis. *Fertil Steril.*, 2013, 99, 690–696.

Kennedy, S. (2003). Genetics of endometriosis: a review of the positional cloning approaches. *Seminars in Productive Medicine.*, 21, 111-118.

Krausz, C; Riera-Escamilla, A. Genetics of male infertility. *Nat Rev Urol.*, 2018, 15(6), 369–384.

Krisher, RL. The effect of oocyte quality on development. *J Anim Sci.*, 2004, 82 E-Suppl, E14–E23.

Kubiszeski, EH; de Medeiros, SF; da Silva Seidel, JA; Barbosa, JS; Galera, MF; Galera, BB. Glutathione S-transferase M1 and T1 gene polymorphisms in Brazilian women with endometriosis. *J Assist Reprod Genet.*, 2015, 32(10), 1531-5.

Kumar, M; Pathak, D; Kriplani, A; Ammini, AC; Talwar, P; Dada, R. Nucleotide variations in mitochondrial DNA and supra-physiological ROS levels in cytogenetically normal cases of premature ovarian insufficiency. *Arch Gynecol Obstet.*, 2010, 282, 695–705.

Lakpour, N; Mirfeizollahi, A; Farivar, S; Akhondi, MM; Hashemi, SB; Amirjannati, N; Heidari-Vala, H; Sadeghi, MR. The association of seminal plasma antioxidant levels and sperm chromatin status with genetic variants of GSTM1 and GSTP1 (Ile105Val and Ala114Val) in infertile men with oligoasthenoteratozoospermia. *Dis Markers.*, 2013, 34(3), 205–210.

Lédée, N; Gridelet, V; Ravet, S; Jouan, C; Gaspard, O; Wenders, F; Thonon, F; Hincourt, N; Dubois, M; Foidart, JM; Munaut, C; Perrier d'Hauterive, S. Impact of follicular G-CSF quantification on subsequent embryo transfer decisions: a proof of concept study. *Hum Reprod.*, 2013, 28(2), 406-13.

Li, H; Zhang, Y. Glutathione S-transferase M1 polymorphism and endometriosis susceptibility: a meta-analysis. *J Gynecol Obstet Biol Reprod.*, 2015, 44(2), 136-44.

Li, X; Pan, J; Liu, Q; Xiong, E; Chen, Z; Zhou, Z; Su, Y; Lu, G. Glutathione S-transferases gene polymorphisms and risk of male idiopathic infertility: a systematic review and meta-analysis. *Mol Biol Rep.*, 2012, 40(3), 2431-8.

Lin, J; Zhang, X; Qian, Y; Ye, Y; Shi, Y; Xu, K; Xu, J. Glutathione S-transferase M1 and T1 genotypes and endometriosis risk: a case-controlled study. *Chin Med J* (Engl)., 2003, 116, 777–780.

Listowsky, I; Rowe, JD; Patskovsky, YV; Tchaikovskaya, T; Shintani, N; Novikova, E; Nieves, E. Human testicular glutathione S-transferases: insights into tissue-specific expression of the diverse subunit classes. *Chem Biol Interact.*, 1998, 111–112, 103–112.

Lu, B; Poirier, C; Gaspar, T; Gratzke, C; Harrison, W; Busija, D; Martin, M; Matzuk Karl-Erik, A; Paul, A; Overbeek, C; Bishop, E. A mutation in the inner mitochondrial membrane peptidase 2-like gene (Immp2l) affects mitochondrial function and impairs fertility in mice. *Biol Reprod.*, 2008, 78, 601–10.

Macanovic, B; Vucetic, M; Jankovic, A; Stancic, A; Buzadzic, B; Garalejic, E; Korac, A; Korac, B; Otasevic, V. Correlation between sperm parameters and protein expression of antioxidative defense enzymes in seminal plasma: a pilot study. *Dis Markers.*, 2015, 436236.

Mancini, A; Meucci, E; Milardi, D; Giacchi, E; Bianchi, A; Pantano, AL; Mordente, A; Martorana, GE; Marinis, L. Seminal antioxidant capacity in pre-and postoperative varicocele. *J Androl.*, 2004, 25(1), 44–49.

Manoj, K; Dhananjay, P; Sundararajan, V; Alka, K; Ammini, AC; Rima, D. *Chromosomal abnormalities & oxidative stress in women with premature ovarian failure (POF).*, 2012, 135(1), 92–97.

Matsuzaka, Y; Kikuti, YY; Goya, K; Suzuki, T. Lack of an association human dioxin detoxification gene polymorphisms with endometriosis in Japanese women: results of a pilot study. *Environ Health Prev Med.*, 2012, 17(6), 512-7.

Meijide, S; Hernández, ML; Navarro, R; Larreategui, Z; Ferrando, M; Ruiz-Sanz, JI; Ruiz-Larrea, MB. Glutathione S-transferase activity in follicular fluid from women undergoing ovarian stimulation: role in maturation. *Free Radical Biology & Medicine.*, 2014, 75(Suppl 1), p. S41.

Mohammadi, M. Oxidative Stress and Polycystic Ovary Syndrome: A Brief Review. *Int J Prev Med.*, 2019, 17, 10, 86.

Moran, C; Tena, G; Moran, S; Ruiz, P; Reyna, R; Duque, X. Prevalence of polycystic ovary syndrome and related disorders in Mexican women. *Gynecol Obstet Invest.*, 2010, 69, 274–80.

Mostafa, T; Rashed, LA; Zeidan, AS; Hosni, A. Glutathione-S-transferase-oxidative stress relationship in the internal spermatic vein blood of infertile men with varicocele. *Andrologia.*, 2015, 47, 47–51.

Mukherjee, SB; Aravinda, S; Gopalakrishnan, B; Nagpal, S; Salunke, DM; Shaha, C. Secretion of glutathione S-transferase isoforms in the seminiferous tubular fluid, tissue distribution and sex steroid binding by rat GSTM1. *Biochem J.*, 1999, 15, 340.

Murri, M; Luque-Ramírez, M; Insenser, M; Ojeda-Ojeda, M; Escobar-Morreale, HF. Circulating markers of oxidative stress and polycystic ovary syndrome (PCOS): A systematic review and meta-analysis. *Human Reprod Update.*, 2013, 19, 268–88.

Nair, RR; Khanna, A; Singh, K. Association of GSTT1 and GSTM1 polymorphisms with early pregnancy loss in an Indian population and a meta-analysis. *Reproductive Bio Medicine Online.*, 2013, 26, 313-322.

Nakahara, K; Saito, H; Saito, T; Ito, M; Ohta, N; Takahashi, T; Hiroi, M. The incidence of apoptotic bodies in membrana granulosa can predict prognosis of ova from patients participating in *in vitro* fertilization programs. *Fertil Steril.*, 1997, 68(2), 312–317.

Nuñez-Calonge, R; Cortés, S; Gutierrez Gonzalez, LM; Kireev, R; Vara, E; Ortega, L; Caballero, P; Rancan, L; Tresguerres, J. Oxidative stress in follicular fluid of young women with low response compared with fertile oocyte donors. *Reprod Biomed Online.*, 2016, 32(4), 446–456.

O'Flynn, O'Brien, KL; Varghese, AC; Agarwal, A. The genetic causes of male factor infertility: a review. *Fertil Steril.*, 2010, 93, 1–12.

Okubo, K; Nagahama, K; Kamoto, T; Okuno, H; Ogawa, O; Nishiyama, H. GSTT1 and GSTM1 polymorphisms are associated with improvement in seminal findings after varicocelectomy. *Fertil Steril.*, 2005, 83, 1579–80.

Olshan, AF; Luben, TJ; Hanley, NM; Perreault, SD; Chan, RL; Herring, AH; Basta, PV; De Marini, DM. Preliminary examination of polymorphisms of GSTM1, GSTT1, and GSTZ1 in relation to semen quality. *Mutat Res.*, 2010, 688, 41–6.

Olszak-Wąsik, K; Bednarska-Czerwińska, A; Olejek, A; Tukiendorf, A. From "Every Day" Hormonal to Oxidative Stress Biomarkers in Blood and Follicular Fluid; to Embryo Quality and Pregnancy Success? *Oxid Med Cell Longev.*, 2019, 1092415.

Polak, G; Koziol-Montewka, M; Gogacz, M; Blaszkowska, I; Kotarski, J. Total antioxidant status of peritoneal fluid in infertile women. *Eur J Obstet Gynecol Reprod Biol.*, 2001, 94(2), 261–3.

Polonikov, AV; Yarosh, SL; Kokhtenko, EV; Starodubova, NI; Pakhomov, SP; Orlova, VS. The functional genotype of glutathione S-transferase T1 gene is strongly associated with increased risk of idiopathic infertility in Russian men. *Fertil Steril.*, 2010, 94, 1144–7.

Ramalhinho, AC; Fonseca-Moutinho, JA; Breitenfeld, L. Glutathione S-transferase M1, T1, and P1 genotypes and breast cancer risk: a study in a Portuguese population. *Mol Cell Biochem.*, 2011, 355(1-2), 265–271.

Ray, SD; Lam, TS; Rotollo, JA; Phadke, S; Patel, C; Dontabhaktuni, A; Mohammad, S; Lee, H; Strika, S; Dobrigowska, A; Bruculeri, C; Chou, A; Patel, S; Pater, R; Manolas, T; Stohs, S. *Oxidative stress is the master operator of drug and chemically-induced programmed and unprogrammed cell death: Implications of natural antioxidants in vivo.*, 2004, 21(1-4), 223-32.

Rier, SE. Environmental immune disruption: a comorbidity factor for reproduction? *Fertility and Sterility.*, 2008, 83, 103-108.

Sabatini, L; Wilson, C; Lower, A; Al-Shawaf, T; Grudzinskas, JG. Superoxide dismutase activity in human follicular fluid after controlled ovarian hyperstimulation in women undergoing *in vitro* fertilization. *Fertil Steril.*, 1999, 72(6), 1027–1034.

Sadeghi, MR. Unexplained infertility, the controversial matter in management of infertile couples. *J Reprod Infertil.*, 2015, 16(1), 1-2.

Safarinejad, MR; Dadkhah, F; Ali Asgari, M; Hosseini, SY; Kolahi, AA; Iran-Pour, E. Glutathione S-transferase polymorphisms (GSTM1, GSTT1, GSTP1) and male factor infertility risk: a pooled analysis of studies. *Urol J.*, 2012, 9, 541–8.

Safarinejad, MR; Shafiei, N; Safarinejad, S. The association of glutathione-S-transferase gene polymorphisms (GSTM1, GSTT1, GSTP1) with idiopathic male infertility. *J Hum Genet.*, 2010, 55, 565–70.

Safarinejad, MR. Infertility among couples in a population-based study in Iran: prevalence and associated risk factors. *Int J Androl.*, 2008, 31, 303–14.

Saleh, RA; Agarwal, A; Sharma, RK; Said, TM; Sikka, SC; Thomas, AJ; Jr. Evaluation of nuclear DNA damage in spermatozoa from infertile men with varicocele. *Fertil Steril.*, 2003, 80(6), 1431–1436.

Salehi, Z; Gholizadeh, L; Vaziri, H; Madani, AH. Analysis of GSTM1, GSTT1, and CYP1A1 in idiopathic male infertility. *Reprod Sci.*, 2012, 19, 81–85.

Schuppe, HC; Wieneke, P; Donat, S; Fritsche, E; Köhn, FM; Abel, J. Xenobiotic metabolism, genetic polymorphisms and male infertility. *Andrologia.*, 2000, 32, 255–262.

Seifati, SM; Parivar, K; Aflatoonian, A; Dehghani Firouzabadi, R; Sheikhha, MH. No association of GSTM1 null polymorphism with endometriosis in women from central and southern Iran. *Iran J Reprod Med.*, 2012, 1, 23-8.

Seino, T; Saito, H; Kaneko, T; Takahashi, T; Kawachiya, S; Kurachi, H. Eight-hydroxy-2'-deoxyguanosine in granulosa cells is correlated with the quality of oocytes and embryos in an *in vitro* fertilization-embryo transfer program. *Fertil Steril.*, 2002, 77(6), 1184–1190.

Shiotani, M; Noda, Y; Narimoto, K; Imai, K; Mori, T; Fujimoto, K; Ogawa, K. Immunohistochemical localization of superoxide dismutase in the human ovary. *Hum Reprod.*, 1991, 6(10), 1349–1353.

Shiou-Sheng Chen, Luke S. Chang, Haw-Wen Chen, Yau-Huei Wei. Polymorphisms of glutathione S-transferase M1 and male infertility in

Taiwanese patients with varicocele. *Human Reproduction.*, 2002, 17, 3, 718–725.

Showell, MG; Mackenzie-Proctor, R; Brown, J; Yazdani, A; Stankiewicz, MT; Hart, RJ. Antioxidants for male subfertility. *Cochrane Database Systematic Reviews.*, 2014, (12).

Storgaard, L; Bonde, JP; Ernst, E; Andersen, CY; Spanô, M; Christensen, K; Petersen, HC; Olsen, J. Genetic and environmental correlates of semen quality: a twin study. *Epidemiology.*, 2006, 17(6), 674–681.

Strange, RC; Spiteri, MA; Ramachandran, S; Fryer, AA. Glutathione-S-transferase family of enzymes. *Mutat Res.*, 2001, 482, 21–26.

Tafuri, S; Ciani, F; Luigi Iorio, E; Esposito, L; Cocchia, N. Reactive oxygen species (ROS) and male fertility. *New Discoveries in Embryology, Bin Wu, Intech Open.*, 2015, 1, 19-40

Tang, K; Xue, W; Xing, Y; Xu, S; Wu, Q; Liu, R; Wang, X; Xing, J. Genetic polymorphisms of glutathione S-transferase M1, T1, and P1, and the assessment of oxidative damage in infertile men with varicoceles from Northwestern China. *J Androl.*, 2012, 33, 257–63.

Tang, M; Wang, S; Wang, W; Cao, Q; Qin, C; Liu, B; Li, P; Zhang, W. The glutathione-S-transferase gene polymorphisms (GSTM1 and GSTT1) and idiopathic male infertility risk: a meta-analysis. *Gene.*, 2012, 511, 218–223

Tesarik, J; Hazout, A; Mendoza, C. Enhancement of embryo developmental potential by a single administration of GnRH agonist at the time of implantation. *Hum Reprod.*, 2004, 19(5), 1176–1180.

Thomas, EJ; Campbell, IG. Evidence that endometriosis behaves in a malignant manner. *Gynecol Obstet Invest.*, 2000, 50 Suppl 1, 2–10.

Tirumala Vani, G; Mukesh, N; Siva Prasad, B; Rama Devi, P; Hema Prasad, M; Usha Rani, P; Pardhanandana Reddy, P. Role of glutathione S-transferase Mu-1 (GSTM1) polymorphism in oligospermic infertile males. *Andrologia.*, 2010, 42(4), 213–217.

Tremellen, K. Oxidative stress and male infertility – A clinical perspective. *Hum Reprod Update.*, 2008, 14, 243–58.

Tsai-Turton, M; Luderer, U. Opposing effects of glutathione depletion and follicle-stimulating hormone on reactive oxygen species and apoptosis in cultured preovulatory rat follicles. *Endocrinology.*, 2006, 147(3), 1224–1236.

Tuo, Y; He, JY; Yan, WJ; Yang, J. Association between CYP19A1, GSTM1, GSTT1, and GSTP1 genetic polymorphisms and the development of endometriosis in a Chinese population. *Genet Mol Res.*, 2016, 19, 15(4).

Turner, TT; Lysiak, JJ. Oxidative stress: a common factor in testicular dysfunction. *J Androl.*, 2008, 29, 488–498.

Uyar, A; Torrealday, S; Seli, E. Cumulus and granulosa cell markers of oocyte and embryo quality. *Fertil Steril.*, 2013, 99(4), 979–997.

Vander Borght, M; Wyns, C. Fertility and infertility: Definition and epidemiology. *Clin Biochem.*, 2018, 62, 2–10.

Velthut, A; Zilmer, M; Zilmer, K; Kaart, T; Karro, H; Salumets, A. Elevated blood plasma antioxidant status is favourable for achieving IVF/ICSI pregnancy. *Reproductive biomedicine online.*, 2014, 75, (Suppl1), S41.

Vichi, S; Medda, E; Ingelido, AM; Ferro, A; Resta, S; Porpora, MG; Abballe A; Nisticò, L; De Felip, E; Gemma, S; Testai, E. Glutathione transferase polymorphisms and risk of endometriosis associated with polychlorinated biphenyls exposure in Italian women: a gene-environment interaction. *Fertil Steril.*, 2012, 97(5), 1143-51.

Wang, YX; Wang, P; Feng, W; Liu, C; Yang, P; Chen, YJ; Sun, L; Sun, Y; Yue, J; Gu, LJ; Zeng, Q; Lu, WQ. Relationships between seminal plasma metals/metalloids and semen quality, sperm apoptosis and DNA integrity. *Environmental Pollution.*, 2017, 224, 224–234.

Whitbread, AK; Tetlow, N; Eyre, HJ; Sutherland, GR; Board, PG. Characterization of the human Omega class glutathione transferase genes and associated polymorphisms. *Pharmacogenetics.*, 2003, 13(3), 131-144.

Wu, QF; Xing, JP; Tang, KF; Xue, W; Liu, M; Sun, JH; Wang, XY; Jin, XJ. Genetic polymorphism of glutathione S-transferase T1 gene and susceptibility to idiopathic azoospermia or oligospermia in northwestern China. *Asian J Androl.*, 2008, 10, 266–70.

Wu, W; Lu, J; Tang, Q; Zhang, S; Yuan, B; Li, J; Di, Wu; Sun, H; Lu, C; Xia, Y; Chen, D; Sha, J; Wang, X. GSTM1 and GSTT1 null polymorphisms and male infertility risk: an updated meta-analysis encompassing 6934 subjects. *Sci Rep.*, 2013, 3, 2258.

Xin, X; Jin, Z; Gu, H; Wu, T; Hua, T; Wang, H. Association between glutathione S-transferase M1/T1 gene polymorphisms and susceptibility to endometriosis, a systematic review and meta-analysis. *Exp Ther Med.*, 2016, 11(5), 1633-46.

Xiong, DK; Chen, HH; Ding, XP; Zhang, SH; Zhang, JH. Association of polymorphisms in glutathione S-transferase genes (GSTM1, GSTT1, GSTP1) with idiopathic azoospermia or oligospermia in Sichuan, China. *Asian J Androl.*, 2015, 17(3), 481–486.

Xu, XB; Liu, SR; Ying, HQ; AZC. Null genotype of GSTM1 and GSTT1 may contribute to susceptibility to male infertility with impaired spermatogenesis in Chinese population. *Biomarkers*, 2013, 18, 151–4.

Ying, HQ; Qi, Y; Pu, XY; Liu, SR; AZC. Association of GSTM1 and GSTT1 genes with the susceptibility to male infertility: result from a meta-analysis. *Genet Test Mol Biomarkers.*, 2013, 17(7), 535-42.

Yu, B; Huang, Z. Variations in Antioxidant Genes and Male Infertility. *Biomed Res. Int.*, 2015, 513196.

Zegers-Hochschild, F; Adamson, GD; Dyer, S; Racowsky, C; Mouzon, J; Sokol, R; Rienzi, L; Sunde, A; Schmidt, L; Cooke, ID; Simpson, JL; Poel, SVD. The International Glossary on Infertility and Fertility Care, 2017. *Hum Reprod.*, 2017, 32(9), 1786–1801.

Zhang, J; Fan, P; Liu, H; Bai, H; Wang, Y; Zhang, F. Apolipoprotein AI and B levels, dyslipidemia and metabolic syndrome in south-west Chinese women with PCOS. *Human Reprod.*, 2012, 27, 2484–93.

Zhao, Y; Marotta, M; Eichler, EE; Eng, C; Tanaka, H. Linkage disequilibrium between two high-frequency deletion polymorphisms: implications for associations studies involving the glutathione-S transferase (GST) genes. *PLoS Genetics.*, 2009, 5(5), e1000472.

Zhu, H; Bao, J; Liu, S; Chen, Q; Shen, H. Null genotypes of GSTM1 and GSTT1 and endometriosis risk: a meta-analysis of 25 case-control studies. *PLoS One.*, 2014, 9(9), e106761.

INDEX

A

acid, ix, 2, 5, 32, 33, 63, 64, 67, 70, 71
active site, 27, 46, 58
activity level, 43
acute lymphoblastic leukemia, 8, 41, 59
acute myeloid leukemia, 12, 42, 50
adenocarcinoma, 38, 42, 49, 50
airway hyperresponsiveness, 12
allele, 6, 8, 9, 11, 12, 13, 34, 46, 99
amino acid, 4, 7, 28, 34, 37, 41, 56, 67, 80, 98, 100
anticancer drug, viii, x, 2, 13, 15, 60, 64
antigen, 8
anti-inflammatory drugs, 10, 12
antioxidant, viii, ix, 14, 26, 37, 44, 49, 61, 63, 65, 66, 68, 70, 73, 87, 91, 92, 101, 107, 109, 111, 112, 114, 115, 117, 120
apoptosis, viii, x, 5, 6, 11, 14, 26, 27, 60, 64, 66, 90, 92, 100, 104, 106, 109, 119, 120
apoptotic pathways, 39
aromatic amines, 75, 86
aromatic hydrocarbons, 36, 75
arsenic, 65, 74, 84, 89
asthma, 10, 11, 15
atherosclerosis, 30, 43, 55
autoimmune hepatitis, 9

B

bacteria, 27, 35, 49, 64
basal cell carcinoma, 12
benign, 42, 48, 49, 57
bioavailability, 13, 105
biological consequences, 53
biological processes, 3
biomarkers, vii, viii, ix, 2, 3, 8, 15, 26, 47, 106
biomolecules, 65
biosynthesis, 5, 29, 67, 90
bladder cancer, vii, x, 42, 51, 64, 65, 75, 76, 77, 78, 79, 80, 81, 82, 83, 84, 85, 86, 87, 89, 90, 91, 92, 93
bladder carcinogenesis, 75, 76
blood, 9, 13, 22, 30, 43, 116, 120
blood plasma, 120
blood transfusion, 9
breast cancer, 7, 13, 40, 61, 89, 117
bronchial epithelial cells, 11
butadiene, 37, 79

C

cancer, iv, v, vii, viii, ix, x, 2, 5, 6, 8, 11, 12, 13, 14, 15, 16, 17, 18, 19, 20, 21, 22, 23, 26, 28, 34, 35, 36, 37, 38, 40, 42, 43, 48, 50, 51, 52, 53, 54, 55, 56, 57, 58, 59, 60, 61, 63, 64, 65, 71, 74, 75, 76, 77, 78, 79, 80, 81, 82, 83, 84, 85, 86, 87, 88, 89, 90, 91, 92, 93, 110, 111, 117
cancer cells, viii, 2, 13, 14, 40
candidates, 10
carcinogen, 40, 75, 76
carcinogenesis, 36, 55, 56, 84, 86
carcinoma, 38, 42, 50, 54, 81, 84, 85
cardiovascular disease, 45
catalytic activity, 5, 11, 48, 80, 82
Caucasian population, 45, 46
Caucasians, 8, 37, 50, 79, 80, 101
cell cycle, 39
cell death, 28, 39, 81, 117
cell differentiation, 81
cell line, 89
cell membranes, 72
cellular detoxification, ix, 63
cervical cancer, ix, 8, 13, 15, 26, 36, 37, 38, 40, 41, 48, 51, 53, 54, 55, 56, 57, 58, 59
chemical, viii, 5, 26, 27, 49, 74, 75, 85, 86
chemotherapeutic agent, 6, 13, 39, 40
chemotherapy, 2, 12, 13, 14, 16, 18, 28, 39, 40, 42, 49, 50, 53, 59, 61, 75, 76
childhood, 40, 50, 59
children, 8, 13, 41, 58
chromosome, 6, 7, 10, 34, 77, 79, 80, 99
chronic renal failure, 9
cigarette smoking, 36, 54, 61, 65, 75, 84
circulation, 103
classes, viii, 2, 3, 5, 26, 27, 28, 34, 36, 74, 81, 98, 115
clinical application, 26
clinical symptoms, 47
cloning, 48, 82, 91, 114
colorectal adenocarcinoma, 54
colorectal cancer, 13, 40, 43, 60, 77, 83, 85
comparative analysis, 56
complex interactions, xi, 96, 97
compounds, ix, x, 3, 5, 6, 8, 11, 15, 29, 30, 31, 39, 41, 63, 66, 68, 69, 70, 72, 73, 76, 95, 98, 104
conjugation, viii, 1, 2, 3, 4, 6, 25, 26, 28, 30, 31, 32, 33, 34, 39, 66, 68, 69, 70, 71, 72, 76, 98, 105
control group, 78, 79, 102
controversial, 77, 79, 117
coronary artery disease, 43, 45, 55
cyclophosphamide, 12, 14, 71
cyclosporine, 14
cysteine, 3, 11, 67, 73
cytochrome, 2, 30, 31, 33, 52, 56, 57, 71
cytoplasm, 67, 73
cytotoxicity, 5, 38, 84

D

deaths, 36, 43, 75
defects, 43
defence, 53, 82, 86, 91
dehydrochlorination, 72
derivatives, 12, 14, 70, 74
detoxification, vii, viii, ix, x, 1, 2, 5, 7, 10, 11, 13, 14, 15, 25, 26, 28, 30, 31, 33, 36, 40, 41, 60, 63, 65, 66, 68, 69, 71, 74, 80, 95, 101, 107, 115
diabetes, 43, 44, 45, 49, 55, 56, 57, 60, 61, 62
diabetic nephropathy, 45, 57
diabetic patients, 43, 57
diabetic retinopathy, 45, 50, 56
disease activity, 110
disease progression, 75
disease susceptibility, iv, vii, ix, 5, 26, 47
diseases, ix, 3, 5, 26, 30, 34, 44, 47, 63, 85, 99, 104, 107

Index

distribution, viii, 26, 27, 78, 79, 87, 116
DNA, vii, ix, 6, 10, 21, 22, 26, 33, 37, 38, 39, 40, 48, 64, 65, 68, 70, 76, 77, 85, 99, 100, 104, 105, 106, 109, 111, 118, 120
DNA damage, ix, 10, 26, 33, 37, 77, 104, 105, 109, 111, 118
DNA repair, 38, 39, 40, 48
drinking water, 89
drug discovery, 3
drug metabolism, 13, 30, 39
drug resistance, 5, 15, 21, 52, 53, 55, 56, 60, 66
drug treatment, 51
drugs, viii, x, 2, 5, 12, 13, 26, 29, 30, 33, 39, 41, 44, 64, 68, 70, 72, 76, 82
dyslipidemia, 121

E

economic development, 43
economic status, 36
encoding, viii, 2, 3, 7, 15, 26, 29, 37, 47, 99, 111, 113
endangered, 99
endometriosis, 97, 104, 110, 111, 112, 113, 114, 115, 118, 119, 120, 121
end-stage renal disease, 84
environment, 36, 65, 120
environmental factors, 42, 65, 75, 100
environmental influences, 99
environmental tobacco, 10
enzymatic activity, viii, 26, 28, 41, 81, 99
enzymes, viii, ix, xi, 2, 3, 5, 7, 10, 14, 15, 25, 26, 28, 33, 34, 35, 36, 37, 39, 40, 51, 53, 55, 56, 60, 63, 66, 69, 70, 71, 72, 73, 74, 75, 76, 81, 84, 85, 87, 90, 92, 93, 96, 98, 107, 111, 113, 115, 119
epithelial cells, x, 11, 64, 66
ethnic groups, ix, xi, 26, 35, 61, 96, 97
ethnicity, 44, 78, 80, 81
ethylene, ix, 5, 8, 64, 79

ethylene oxide, ix, 8, 64, 79
etiology, xi, 65, 76, 96, 98, 104
evidence, xi, 48, 65, 75, 86, 93, 96, 97
excretion, x, 39, 64, 66, 76
exposure, 10, 11, 35, 37, 65, 72, 74, 75, 76, 86, 89, 105, 120
expression of GSTs, x, 64

F

fallopian tubes, 103
families, viii, ix, 3, 26, 27, 63
family members, 76, 98
fertility, 101, 103, 112, 115, 119
fertilization, 100, 106, 116, 117, 118
fluid, 54, 97, 101, 106, 116, 117
follicle, 105, 106, 107, 119
follicular fluid, 106, 107, 114, 115, 116, 117
formation, x, 4, 6, 30, 34, 39, 64, 66, 67, 69, 99, 105
free radicals, 17, 38, 64, 70, 91, 109

G

gastrointestinal tract, 42, 71
gene expression, x, 6, 33, 96, 105, 106
gene polymorphisms, 15, 21, 48, 49, 53, 61, 62, 65, 81, 92, 102, 110, 111, 113, 114, 115, 118, 119, 121
gene regulation, 86
genes, viii, ix, 2, 3, 5, 7, 8, 15, 26, 29, 34, 35, 37, 39, 40, 42, 44, 46, 47, 48, 49, 51, 55, 56, 59, 63, 73, 74, 75, 76, 85, 88, 90, 91, 93, 97, 98, 99, 102, 104, 105, 110, 111, 113, 120, 121
genetic alteration, 15
genetic factors, xi, 65, 76, 96, 97
genetic marker, 112
genetic predisposition, 75
genetic variants, iv, vii, viii, xi, 26, 36, 44, 47, 82, 96, 114

genomic integrity, 76, 109
genotype, 7, 8, 9, 10, 12, 14, 37, 38, 40, 41, 45, 50, 54, 77, 78, 79, 80, 81, 87, 101, 102, 104, 105, 108, 109, 111, 117, 121
germ cells, 100, 101, 105, 110
germline polymorphisms, 54
glucocorticoid receptor, 6
glutathione (GSH), iv, v, vii, viii, ix, x, 1, 2, 3, 4, 5, 6, 10, 15, 16, 17, 18, 19, 20, 21, 22, 23, 25, 26, 27, 28, 29, 30, 31, 32, 33, 34, 36, 39, 40, 42, 46, 48, 49, 50, 51, 52, 53, 54, 55, 56, 57, 58, 59, 60, 61, 62, 63, 65, 66, 67, 68, 69, 70, 71, 72, 73, 74, 75, 76, 81, 82, 83, 84, 85, 86, 87, 88, 89, 90, 91, 92, 93, 95, 98, 107, 108, 109, 110, 111, 112, 113, 114, 115, 116, 117, 118, 119, 120, 121
glutathione disulfide (GSSG), 34, 67, 68
glutathione-S-transferases (GSTs), viii, ix, x, 2, 3, 4, 5, 6, 10, 13, 14, 15, 16, 17, 25, 26, 27, 28, 34, 35, 36, 37, 39, 40, 41, 42, 43, 54, 56, 61, 63, 66, 69, 70, 71, 72, 74, 75, 77, 81, 91, 95, 96, 98, 99, 101, 104, 105, 106, 107, 111
glycine, 3, 33, 65, 67
GSH conjugates, 34, 67, 68, 70
GST, iv, vii, viii, ix, x, 1, 2, 3, 4, 5, 6, 7, 9, 10, 11, 12, 13, 14, 15, 17, 25, 26, 27, 34, 36, 37, 38, 39, 40, 41, 42, 43, 44, 47, 53, 54, 58, 59, 61, 64, 65, 66, 67, 68, 69, 70, 71, 72, 73, 76, 81, 86, 89, 93, 98, 100, 102, 103, 106, 107, 108, 110, 121
GST enzymes, 66, 71, 72, 73, 81, 107
GSTM1, ix, 7, 8, 9, 10, 13, 16, 18, 20, 21, 22, 26, 30, 36, 37, 38, 40, 41, 42, 43, 44, 45, 46, 48, 49, 50, 51, 52, 53, 54, 55, 56, 57, 59, 61, 76, 77, 78, 81, 82, 86, 87, 88, 89, 92, 99, 101, 102, 104, 105, 108, 109, 110, 111, 113, 114, 116, 117, 118, 119, 120, 121
GSTM1-null genotype, 42, 78

GSTP1, ix, 8, 9, 10, 11, 12, 13, 14, 18, 20, 21, 22, 26, 31, 34, 36, 37, 38, 40, 41, 42, 43, 46, 48, 52, 54, 55, 56, 57, 59, 61, 76, 77, 80, 81, 82, 85, 87, 89, 92, 99, 102, 104, 108, 113, 114, 118, 120, 121
GSTP1 polymorphism, 21, 42, 80, 81, 82, 104
GSTT1 null genotype, 8, 9, 10, 41, 45, 50, 79, 80, 99, 101, 102, 104

H

harmful, 39, 70, 72, 74
hepatotoxicity, viii, 2, 7, 15
homeostasis, viii, 26, 27, 73, 74
hormones, 29, 55, 76, 110
human, vii, ix, 4, 5, 9, 14, 15, 34, 48, 50, 52, 54, 56, 57, 63, 76, 81, 82, 85, 88, 89, 91, 99, 106, 107, 109, 111, 112, 113, 115, 117, 118, 120
human body, 99, 107
human leukocyte antigen, 9
human papilloma virus, 57
hydrogen, ix, 64, 92
hydrogen peroxide, ix, 64, 92
hydroperoxides, 5, 28, 32, 34, 68, 71, 74
hyperandrogenism, 97, 112
hypermethylation, 8, 12
hypothesis, 10, 14, 41, 81, 105

I

idiopathic, 98, 102, 109, 112, 115, 117, 118, 119, 120, 121
immune response, 11, 14
immunomodulatory, 36
in utero, 10
in vitro, 116, 117, 118
in vivo, 117
incidence, xi, 15, 54, 55, 83, 87, 96, 97, 100, 116

Index

India, 22, 25, 47, 48, 55, 59, 108
Indians, 60, 90
individuals, 6, 7, 8, 9, 10, 11, 34, 37, 38, 41, 43, 44, 45, 78, 80, 87, 102
induction, 60, 74, 76, 84, 90, 93
industrial chemicals, 5
infection, 36, 52, 55, 100
infertility, iv, v, vii, xi, 95, 96, 97, 98, 99, 100, 101, 102, 103, 104, 105, 107, 109, 110, 111, 112, 113, 114, 115, 116, 117, 118, 119, 120, 121
inflammation, ix, 11, 26, 32, 55
inflammatory disease, 10
inhibition, 6, 14, 56, 105
insecticide, 72, 88, 91
insects, 35, 64, 72, 82, 86
insulin, 43, 44, 112
insulin resistance, 43, 44
integrity, 76, 100, 109, 120
isoleucine, 37, 41, 46, 80

L

leukemia, 12, 13, 40, 41, 76
leukotrienes, 5, 11, 32, 68, 70
lipid peroxidation, 6, 33, 70, 71, 73, 92, 105
lipid peroxides, 68
lipids, vii, viii, x, 25, 27, 43, 64, 65
lipoproteins, viii, 25, 27
liver, 3, 6, 7, 9, 10, 14, 71, 88, 99
liver transplant, 9
liver transplantation, 9
locus, 34, 35, 60, 88, 99
lung cancer, 37, 59, 60, 61, 83
lymphoma, 41, 58, 81

M

malaria, 7, 14, 84, 88
mammals, 49, 60, 64, 70
matrix metalloproteinase, 91

medical care, 110
medulloblastoma, 58
mellitus, ix, 26, 34, 43, 49, 52, 55, 56, 57, 59, 61, 62
meta-analysis, 51, 77, 78, 79, 80, 81, 82, 84, 85, 86, 92, 101, 103, 109, 111, 114, 115, 116, 119, 121
metabolism, 2, 3, 6, 7, 10, 30, 31, 32, 35, 43, 68, 82, 84, 91, 105, 118
metabolites, viii, 1, 4, 6, 7, 8, 10, 13, 14, 15, 26, 27, 33, 36, 43, 49, 69, 70
metabolizing, viii, 25, 26, 33, 34, 36, 76, 107
microorganisms, vii, viii, 1, 2, 3, 5
mitochondria, 64, 105
mitochondrial DNA, 105, 114
molecules, ix, 2, 38, 40, 63, 68, 73, 76, 113
multicellular organisms, 27
multiple sclerosis, 10
mutagenesis, x, 64
mutation, 7, 115
mutations, 33, 37, 76, 99

N

neurodegeneration, 51
neurodegenerative diseases, 10, 82
null, 7, 8, 9, 10, 13, 34, 35, 37, 40, 41, 42, 44, 45, 50, 55, 77, 78, 79, 80, 82, 84, 87, 89, 99, 101, 102, 104, 105, 109, 110, 111, 118, 121

O

occupational exposure, 65, 75, 86
oligospermia, 120, 121
oligozoospermia, 97
oocyte, 100, 104, 105, 106, 107, 113, 114, 116, 120
oogenesis, 105, 107
ovarian cancer, 8, 40, 54, 59, 110

ovarian failure, 97, 111, 112, 115
ovarian tumor, 48
ovaries, 42, 98, 103, 105, 109
overexpression of GSTs, x, 64
oxidation, 2, 65, 66, 70, 73, 74
oxidation products, 70
oxidative damage, 39, 40, 89, 102, 109, 119
oxidative stress, viii, ix, x, 1, 5, 7, 13, 25, 26, 29, 30, 34, 43, 52, 53, 62, 63, 64, 65, 72, 73, 74, 75, 82, 84, 90, 95, 96, 97, 98, 99, 100, 101, 103, 104, 105, 106, 107, 108, 109, 110, 111, 113, 114, 115, 116, 117, 119, 120
oxygen, 39, 40, 64, 68, 82, 88, 100, 113, 119

P

parasites, viii, 2, 14, 15, 91
pathogenesis, xi, 75, 76, 86, 96, 97, 104
pathophysiology, xi, 85, 96, 97, 105
pathways, 26, 33, 36, 39, 44, 51, 56, 67, 76, 104
peroxidation, viii, 6, 25, 27, 29, 70, 71
peroxynitrite, ix, 64, 68
pharmacology, 110, 113
phase II detoxification enzymes, 28, 69
phase II metabolic enzymes, ix, 63
plants, vii, viii, 1, 2, 27, 35, 49, 64, 72
plasma membrane, 100, 113
platinum, 13, 14, 38, 39, 40, 41, 76, 89
platinum base chemotherapy drugs, 76
polychlorinated biphenyl, 120
polycyclic aromatic hydrocarbon, ix, 7, 64, 71, 75, 83
polycyclic aromatic hydrocarbons, 7, 36, 71, 75, 83
polycystic ovarian syndrome, 112
polymorphism, 2, 6, 8, 11, 12, 16, 17, 19, 20, 22, 23, 37, 41, 42, 44, 45, 46, 51, 53, 54, 57, 58, 60, 61, 77, 78, 79, 80, 81, 82, 85, 86, 89, 91, 92, 93, 100, 109, 110, 112, 113, 114, 118, 119, 120
population, ix, 6, 8, 9, 10, 12, 26, 34, 37, 45, 46, 49, 50, 52, 55, 56, 78, 79, 80, 88, 91, 99, 101, 103, 113, 116, 117, 118, 120, 121
pregnancy, vii, xi, 9, 10, 96, 106, 107, 116, 120
prognosis, 3, 8, 12, 39, 116
proliferation, viii, 5, 11, 26, 27, 62, 81, 104
pro-oxidant, 65, 66, 74, 82
prostaglandin, 32, 70
prostaglandins, 5, 11, 33, 68
prostate cancer, 5, 6, 8, 12
prostate gland, 99
protection, viii, 25, 34, 40, 50, 53, 66, 69, 71, 76, 90, 92, 93, 111, 113
protective role, 12, 101
protein structure, 4
protein synthesis, 105
proteins, vii, x, 1, 3, 4, 11, 40, 48, 64, 65, 68, 69, 72, 73, 76, 81, 82, 104, 105
proto-oncogene, 89
pulmonary diseases, ix, 26

R

radiotherapy, 8, 12, 39, 40
reactions, viii, x, 1, 2, 5, 15, 29, 34, 64, 66, 68, 69, 70, 71, 72
reactive nitrogen species (RNS), viii, 25, 27, 66
reactive oxygen species (ROS), iv, vii, viii, ix, x, 10, 25, 27, 29, 39, 40, 60, 63, 64, 65, 66, 68, 82, 87, 95, 96, 98, 100, 103, 104, 105, 106, 107, 109, 113, 114, 119
reactive species (RS), iv, vii, ix, 64, 66, 67, 68, 69, 70
recurrence, 8, 15, 38, 41, 75, 91
reproduction, 104, 108, 109, 117

resistance, 2, 5, 13, 14, 15, 26, 28, 39, 40, 43, 44, 49, 52, 53, 55, 56, 58, 59, 60, 66, 67, 71, 72, 73, 76, 87, 88, 91
response, ix, 13, 26, 35, 39, 41, 42, 44, 48, 50, 84, 103, 112, 116
risk, vii, viii, ix, 2, 6, 7, 8, 9, 10, 11, 12, 14, 26, 34, 35, 36, 37, 38, 41, 42, 43, 44, 45, 46, 47, 49, 50, 51, 52, 54, 55, 56, 57, 59, 60, 61, 65, 75, 76, 77, 78, 79, 80, 81, 83, 84, 85, 86, 88, 89, 90, 91, 92, 93, 97, 101, 102, 104, 107, 111, 113, 114, 115, 117, 118, 119, 120, 121
risk assessment, vii, ix, 26, 47
risk factors, 36, 75, 85, 118
risk of bladder cancer, 75, 76, 77, 78, 79, 80, 81

S

semen, 100, 117, 119, 120
seminal vesicle, 99
senescence, viii, 26, 27
sexual intercourse, vii, xi, 96
signaling pathway, viii, 2, 5, 11, 69, 74, 81, 90
single nucleotide polymorphisms, x, 44, 50, 64
smoking, 10, 36, 48, 49, 52, 55, 57, 59, 60, 80, 83, 84, 86, 87
smoking cessation, 60
smooth muscle, 62, 84, 93
smooth muscle cells, 84, 93
species, vii, ix, x, 10, 29, 34, 39, 40, 49, 60, 63, 64, 66, 67, 81, 82, 87, 95, 100, 109, 113, 119
sperm, 97, 100, 101, 102, 109, 110, 114, 115, 120
spermatogenesis, 100, 121
squamous cell, 38, 43, 50
squamous cell carcinoma, 38, 43, 50

stress, x, 39, 44, 52, 90, 93, 95, 99, 101, 103, 105, 109, 110, 113, 114, 116, 117, 119, 120
structure, ix, 3, 4, 28, 58, 63
substrate, viii, 4, 6, 10, 11, 25, 28, 33, 37, 56, 70, 72, 108
substrates, 2, 4, 5, 6, 27, 28, 39, 67, 69, 71, 72, 73, 82, 108
superfamilies, ix, 63, 66
survival, ix, x, 13, 26, 38, 39, 40, 41, 51, 59, 60, 64, 69, 84, 99
susceptibility, vii, viii, ix, 2, 5, 7, 12, 16, 17, 22, 23, 26, 36, 37, 42, 44, 47, 49, 53, 54, 57, 60, 61, 66, 77, 78, 79, 80, 81, 82, 85, 87, 89, 90, 91, 92, 93, 99, 102, 108, 109, 110, 112, 114, 120, 121
syndrome, 97, 111, 112, 116
synthesis, viii, 2, 58, 104

T

target, viii, 2, 14, 29, 40, 72, 82
testis, 30, 54, 99, 100, 101, 109
therapeutic agents, 14
therapy, 12, 15, 39, 55, 87
thermostability, 108
thyroid gland, 99
tissue, 9, 10, 38, 43, 52, 97, 100, 112, 115, 116
tissue engineering, 112
tobacco, ix, 10, 26, 36, 37, 42, 64, 65, 76
tobacco smoke, ix, 26, 36, 37, 64, 65
tobacco smoking, 10, 36, 76
toxic effect, 65
toxic xenobiotics, 34, 65, 68
toxicity, ix, 26, 35, 41, 44, 61, 87, 90, 98, 100
toxicology, 52, 87, 110, 113
transcription, 6, 29, 34, 105, 106
transcription factors, 29, 34, 105, 106

treatment, vii, ix, x, 9, 12, 13, 15, 26, 38, 39, 40, 41, 44, 47, 48, 58, 64, 66, 75, 109, 110
tumor, x, 13, 36, 39, 40, 52, 60, 64, 66
tumorigenesis, x, 64
tumors, 8, 41, 53, 66, 81, 82
type 2 diabetes, ix, 26, 34, 49, 50, 52, 53, 57, 59, 61
type 2 diabetes mellitus, ix, 26, 34, 43, 44, 49, 52, 55, 59, 61

U

urinary bladder, 81, 88, 89, 90, 93
urinary bladder cancer, 81, 89, 90
urinary tract, x, 64
urine, x, 64, 66, 77

W

water, viii, 25, 31, 33, 39, 69, 70
worldwide, 36, 43, 74, 83, 96, 100

X

xenobiotics, ix, 2, 30, 31, 33, 35, 55, 63, 65, 68, 69, 70, 72, 74, 76, 86, 87, 98, 100, 106

Y

Y chromosome, 109

Z

zygote, 106